Dance of the Continents

Also by John W. Harrington:

To See a World
Discovering Science

DANCE
of the
Continents
Adventures with Rocks and Time
JOHN W. HARRINGTON

J. P. TARCHER, INC.
Los Angeles

Distributed by Houghton Mifflin Company
Boston

Library of Congress Cataloging in Publication Data

Harrington, John Wilbur, 1918–
Dance of the continents.

Bibliography: p. 233
Includes index.
1. Geology—Philosophy.   2. Science—Philosophy.
I. Title.
QE6.H37   1983   550'.1   82–16979
ISBN 0–87477–168–4
ISBN 0–87477–247–8 (ppbk.)

Requests for such permissions should be addressed to:

J. P. Tarcher, Inc.
9110 Sunset Blvd.
Los Angeles, CA 90069

Design by John Brogna/Mike Yazzolino
Illustrated by Adrienne Picci

Manufactured in the United States of America

V   10 9 8 7 6 5 4 3 2 1
First Edition

To
Emma and Alice

Contents

"To see a World in a grain of sand,
And a Heaven in a wildflower,
Hold Infinity in the palm of your hand,
And Eternity in an hour."

William Blake
"The Auguries of Innocence"

Follow Me

Perhaps, But who is this Pied Piper
Beckoning from the mouth of Time's
Tunnel?

I'm a minstrel, a teacher. I sing of ancient stories told by rocks, rivers, mountains, and plains. Follow me, and I'll show you how to escape from a culture-bound preoccupation with present time. I've come to lead you toward a wondrous adventure with the realities of *all time*. I promise joy, a special kind of joy that sweeps over us when we form new insights or discover that we have learned to understand something completely. William Blake felt this kind of exhilaration when he wrote:

> To hold Infinity in the palm of your hand
> And Eternity In an hour.

We must become acquainted in order to work together as a team. You need to know a little about me, and I need to explain the path we'll follow throughout the book. In order to enable you to glimpse the exciting "real" world outside your own windows, a period of apprenticeship must be completed. Then you will approach a world that you'll be able to see and understand for yourself.

This book has an odd history. Many years ago, when I was a graduate student mapping rocks south of Chapel Hill, North Carolina, I was caught in a sudden thunderstorm. There were two places for shelter: an abandoned barn or a run-down, but invitingly humming, two-room schoolhouse. I was bashful and rode my motorbike into the barn, knowing full well that the schoolchildren, bored on a rainy day, would have been overjoyed to have heard stories about the earth from an enthusiastic stranger. That choice of a dry, musty barn was a mistake. I should have chosen that roomful of eager children. That experience convinced me that I should share my enthusiasms whenever the proper opportunity arose, as it did when I was asked to write this book.

Selecting material was a difficult task. Fortunately, editors have sharp razors to trim the unnecessary threads that fail to advance the story. Our book has developed into an extended field trip to many parts of the earth, led by some of the greatest geologists of all time. It illustrates their joys in discovering the larger contexts containing seemingly minor observations.

My editor, Janice Gallagher, made an unusually imaginative demand as soon as we began to work together: "Write the book around a single, unifying law of science." That obviously was a good idea, but I didn't know any laws that would do. Her response to that situation was remarkable: "Make one up." So we did exactly that, and here it is, "Harrington's First Law of Science": "Nature is scrutable when everything is seen in context." Thereupon, the book became a study in context.

A book is a device for exchanging ideas. The flow of thought from point to point is easier to follow if a reader can anticipate the developing order of things. Our study of context begins, as it should, in the field, watching geologists at work, seeing them caught up in the lure of the hunt. Once we know what geologists do, we need to learn how they think. We'll know we're making progress when we discover that we have learned to read rocks and see time as they do. Then nothing will seem beyond us. We will savor the intellectual daring of geologists who lived over 2,000 years ago. We will see how a time scale covering the last 4.5 billion years was assembled and calibrated. We will even begin to see the structures produced when continents break apart and then collide with one another. Our apprenticeship will end as we develop the ability to see the real world for ourselves and experience the thrill of knowing and of knowing that we know.

Early success in developing this ability is important, for it becomes a tremendous incentive to further achievement. A combination of imagination and reason can lead to remarkable insights. We'll use the iron in a common nail to demonstrate the importance of viewpoint. A scientist can read the story of creation in the history of that iron and of every other heavy chemical element, including the carbon and oxygen in our own bodies.

All of the complex elements are composed of hydrogen and wandering subatomic particles that must have been assembled under the intense pressures found only in the interior of stars. The heavy elements present on earth could not have escaped from the sun. Therefore, when we view an iron nail in this way, we are seeing at least six billion years back in time and into the interior of a star, or stars, far older than our sun. Our solar system is made of recycled materials that were once "burped" out into space during a stellar explosion and collected again into the form we know, some 4.5 billion years ago. Albert Einstein put this all very nicely in a single sentence!

"The most incomprehensible thing about the Universe is that it is comprehensible."

No wonder geologists trust reason rather than authority and think of creation as a continuing adventure. There is your springboard. Dive in!

Ab clave ad astra, addicens tuus coepit.
("From a nail to the stars, your apprenticing has begun.")

Acknowledgments

Creative achievements seem to be spurred by a critical mass of thinkers pushing and pulling one another along from idea to idea. Something may be said in jest around a coffee table which will reappear months later in a new setting as an important insight. The task of giving credit for particular contributions is a difficult one. It's hard to remember which colleague or student first expressed a viewpoint that, with a little twist, became a part of me. I can only thank my friends en masse, with the hope that they will savor their ideas as much on reading as they did in proclamation.

The work of five of my own great teachers is reflected throughout this book. Miss Elizabeth Bently Moon at Northside Junior High in Richmond, Virginia, was vibrant with the thought of a "ring of fire" circling the Pacific Ocean. She taught me that geology is an exciting science. Dr. Roy Jay Holden at Virginia Tech taught me that geology is a romantic science. Dr. Jasper L. Stuckey of North Carolina State University offered a technically hopeless student the opportunity to go to graduate school. Dr. Gerald R. MacCarthy at the University of North Carolina taught me to look for the larger contexts. Dr. Claude C. Albritton, Jr., taught me to think. These are all very good things to do for someone.

Specific credits go, first, to my wife, Emma, who typed many versions of the manuscript and shared a number of the field experiences. My sister Alice, master of the comma, was our arbiter of grammar, clause, and effect. Martha Wharton, Lady d'Artagnan of reference librarians, was never at a loss to find the facts. Mark

Olencki, graphic artist, made magnificent photo prints from nearly useless color slides. His gift of time and ingenuity is difficult to acknowledge properly. Professor John L. Salmon offered unstinting support in all matters of foreign words and phrases. Many geologists in various parts of the world contributed technical facts and wisdom whenever called upon. They know their service and I love them dearly for giving help so freely.

My editors, Janice Gallagher and Katherine Leiner, are responsible for turning an often soft Southern homily into an accurate cosmopolitan adventure. Bless their patience.

John W. Harrington
Professor of Geology Emeritus
Wofford College
Spartanburg, South Carolina

PART 1

Geology
and
Geologists

Eighteenth-century cartoon of Dr. James Hutton.

ONE

The Lure
of the Hunt

"We were hunters before we were farmers."

HENRY S. JOHNSON, JR., CONSULTING GEOLOGIST
Personal Communication

W e were hunters and gatherers for more than five million years before learning to farm successfully. A gene-imprinted lust to follow the old game trails yet may stir within us all. If so, field geologists are certainly members of the breed for they are a curiously driven set of explorers, trackers, hunters, and gatherers. We can learn a great deal about the science of geology by watching a few of them at work. Anticipation, derring-do, and the intellectual and physical fun of it all: traits that belong to sportsmen and artists characterize scientists as well. It's sometimes difficult to distinguish their work from their play, or hardship from adventure. Geologists tend to romanticize things. A pocketful of peanuts and raisins for lunch becomes a picnic, and tavern talk in the evening, a university. Geologists seem to live in the hunter's state of expectant hope: who knows what surprise may turn up at any moment?

On the day of Queen Elizabeth's Silver Jubilee, I was at Lyme Regis, a small town on the southern coast of England, the setting for John Fowles's best-selling novel, *The French Lieutenant's Woman.*

Every amateur fossil collector in Great Britain appeared to be on the cliffs that day, searching in the crumbling, dark limestones and shales for 140 million-year-old marine fossils of Jurassic age. It was hard for me to imagine that anything of value could have been overlooked by this mob of hunters and gatherers. Nevertheless, I had come to Lyme Regis to sample its treasures and was determined not to leave empty-handed.

I found the beach to be a great mass of broken pieces of chert and flint, worn from the cliffs along various parts of the coastline. These are ideal materials from which to make stone tools and weapons. Primitive man used them extensively. Here was an unexpected opportunity to follow the trail of the ancient Britons.

I began to search among the boulders clustered on the shore. Perhaps through the years they had offered enough protection from harassing storm waves and high tides to have preserved something for me. In less than three minutes I had found half of an ancient, but recently broken, stone axhead wedged between the rocks. Its edges were a little dulled by wave wear; yet the delicate chipping that defined the ax as the work of a human was undeniable. Weathering had changed the outside texture and covered the older surfaces with a soft, porous patina. Newly made breaks were distinguished by exposures of fresh, unweathered rock.

I was less impressed by the technical factors of the ax's survival in the violent beach environment than I was by the human factors of its creation. Survival on the beach was simply a matter of sufficient protection from the mechanical and chemical agents that can destroy rocks; explaining that sort of thing is routine for a geologist. The more exciting questions dealt with people: who made the ax and when was it done? The work was obviously that of a craftsperson. Did he or she share the same kinds of aesthetic pleasures that I was experiencing millennia later?

Heraclitus, a Greek philosopher of the fifth century B.C., said that no man could step into the same river twice. That's true in one way, yet my Stone Age friend had seen, heard, and smelled the same part of the English Channel that I was experiencing. Our hands had touched the same piece of stone. We were both hunters who thrilled with the anticipation of enriching our lives by looking in the best places for the best game. Separation in time was all that kept us from sharing my peanuts and raisins as well as some of our thoughts about the world and its ways. For me, Jubilee Day was spent in the outer fringes of reality somewhere between science and romance. Surely, this is the way William Blake felt when he wrote of holding infinity in the palm of his hand.

Experiences of this sort are quite common in field-oriented sciences, where a practitioner is often alone with his or her thoughts. Years ago when I was still an undergraduate studying mining engineering, I spent a happy Sunday afternoon in the Black Heath of Chesterfield County near Richmond, Virginia. I had gone there to find plant fossils in the shales of the Richmond Triassic Basin. The rocks are about 180 million years old and contain a few low-grade coal seams of submarginal economic value. Optimistic entrepreneurs had initiated sporadic attempts to mine the area since colonial times. After poking my head into three or four eighteenth-century mine tunnels exposed in the walls of a more recently cut open pit, I decided to split open a block of black shale with the chisel edge of my geologist's pick. I chipped carefully along the line of a bedding plane and watched a crack open completely around the block. Suddenly, the halves fell apart revealing a magnificent, bladelike leaf about fifteen inches long and an inch and a half wide. Fossils are usually seen as casts or molds of stony material that follow the shape of some plant or animal that had been entombed while the rock was being deposited. This one was different. The actual leaf lay before me as a thin, mechanically free film of carbon. Every vein and pore was there. For a moment the leaf was perfect. Suddenly the breeze caught the edge of the leaf and began to break it into brittle, jagged fragments. I thought of the original green leaf that had waved so easily in the winds of Triassic age before it was shut off from sunlight by death and burial. What to do? For a moment I was paralyzed by the enormity of my responsibility. Suddenly, I had an unusual flash of genius, and a sense of proportion: the world's finest

Triassic salad was going to waste. I couldn't let that happen . . . so I ate it!

Eating the data may be all right for students, but working geologists must be pragmatists. They are paid to deal with tangible things in a logical manner. Yet even in commercial ventures there is an element of romance. Exploration for phosphate deposits beneath concealing sands of the coastal plain of the southeastern states may lead to better nutrition and a better life for millions of people in Asia and Africa, since phosphate fertilizers improve soil productivity in areas of marginal agricultural potential. The chain of relationships linking geological prospecting for minerals and the quality of life available to all people who live in an industrial society is unbroken.

Back in the 1970s, Henry Johnson, a farsighted, consulting geologist based in Charleston, South Carolina, had this experience while working at the base of the food chain.

His assignment was to create new wealth by discovering phosphate deposits that could be mined and sold at profit. Johnson's method consisted of drilling holes in strategic places and studying core samples taken at various depths. The plan was comparable to hunting for apples in an unknown country by searching the wastelands to find an orchard, then a tree, and eventually the apple. This is a method of homing in on success by eliminating places where failure has been experienced. On the last day of the scheduled field season Johnson reacted intuitively as a gambler.

Years earlier a driller had reported encountering about fifty feet of phosphatic sand in a water well many miles away from Johnson's prospecting area. The report indicated that the deposit, even if properly identified, was too deep to be mined successfully. Johnson's idea was, first, to drill a new hole beside the water well to verify the report and, then, to try to locate a place where mining could be done profitably.

Things went badly from the start. Technical difficulties forced Johnson's driller to give up the first test hole and move the rig to a new location. Wild thunderstorms, encircling the rig, threatened the work with glowering, black clouds, heavy rains, and spectacular displays of lightning. Late in the afternoon the driller finally reached the proper depth and brought up a core from the critical section. Johnson and his assistant waited at the core barrel as the rock slid from its steel case. Their pulses raced in suspense.

Twenty feet of blackish-brown fish-roe-sized pellets of concentrated phosphate emerged. It seemed as pure as fine caviar! In order to see the interior of the core, Johnson split the soft, uncemented rock with his heavy butcher knife. There was no error: it was solid phosphate. The two geologists looked up simultaneously, then their heads turned in unison as they stared around the full circle of the horizon. Johnson broke the silence.

"By God, we've found the scent of the bloody meat! There is a phosphate factory out there somewhere! We're going to shut this rig down until we can figure out where to find it."

That's exactly what they did: paused, evaluated their information, and eventually discovered many millions of tons of mineable phosphate ore.

Field geologists function as detectives. The acts of nature usually have been completed long before they arrive on the scene. Their task becomes one of reconstructing the events that produced the earth as they found it. Their technique is to start with the most recent event and work back in time. Sherlock Holmes called this method, thinking backward. Henry Johnson's phosphate hunt offers a beautiful example.

He began with the knowledge that phosphate pebbles are round because they have been rolled and worn by ocean currents drifting across the sea floor. This meant that a linear current system must have been in operation between the site of the natural phosphate "factory" and the drill hole where Henry "picked up the scent." Johnson also knew that phosphate rocks are marine sediments that were laid down as part of the submerged apron of sand geologists call the continental shelf. Using the pattern of contemporary ocean currents on the continental shelf as a guide, it was an easy matter to predict the best direction in which to explore.

The natural phosphate factory site was located under a thin

cover of younger sediments. Originally it had been a submerged hill where the physical conditions were just right for the chemical precipitation of phosphate pellets from seawater. The ore body was waiting for him just below the crest of the hill. It's easy to imagine the triumphant joy Johnson felt when the shape, thickness, and value of the ore body were defined by further drilling.

Geological fieldwork is a hunt for meaning. Rocks, minerals, fossils, and structures are just intermediate steps, missing links that must be identified before ignorance gives way to understanding. We've already seen a good example of this kind of thinking in the case of the iron nail that was used to introduce an expanded concept of creation: some small observation suddenly expands into a wonder story. Such was Edwin Colbert's 1969 adventure with a fossil tooth.

The story actually begins a decade earlier when explorers from New Zealand found thick coal seams exposed above the glacial ice 8,500 feet up in the central Trans-Antarctic Mountains. They called the place Coalsack Bluff because it curved inward in the form of a cul-de-sac. Several years later Peter J. Barrett, a geologist from New Zealand, found part of the lower jaw of an amphibian enclosed with ancient stream gravels on Graphite Peak, approximately seventy-five miles from Coalsack Bluff. The bone was brought for identification to Dr. Colbert, then curator of fossil reptiles and amphibians at the American Museum of Natural History in New York. This, he concluded, was the first ancient fossil bone ever found anywhere on the continent.

One bone is enough to imply an entire, breeding herd and a balanced ecological system by the coal plants that took their energy from the sun. Coal plants also imply a reasonably warm, humid climate. The sedimentary rocks of the central Trans-Antarctic Mountains obviously were going to be a great place to look for fossil bones of animals capable of having lived under these conditions. Barrett's amphibian jaw fragment was sufficient proof to justify an expanded effort to find the facts and discover their meaning. An expedition, led by Dr. Colbert (by then retired from his post in New York and active as curator of vertebrate paleontology at the Northern Arizona

Museum in Flagstaff), was set up to take advantage of the Antarctic summer of 1969–70. (Logistical support, including food, transportation, and housing, was provided by the United States Navy.) Why was a man who had retired already from a full life's work driven to this arduous task? Perhaps the answer lies in Rudyard Kipling's line: "The bleating of the lamb excites the tiger."

Coalsack Bluff became an exploration target more by accident than by plan. The flat ice field adjacent to it was ideal for a support base. John Splettstoesser of the Institute of Polar Studies at Ohio State University and a crew of navy seabees moved in first to establish the camp and erect four prefabricated Quonset huts. While they were there, Splettstoesser and two seabees took a motor-toboggan ride to the Coalsack to look around. They discovered fossil *Glossopteris* leaves, a tongue fern characteristic of the Permian Period some 250 million years ago. By the time the geological team arrived at camp, they found themselves grounded by weather with little to do but examine the local rocks at Coalsack Bluff. They were back by noon with surprising news: vertebrate fossils and a happy hunting ground.

Here is part of Dr. Colbert's first report to a colleague still working at the museum in New York:

Dear Bobb:

We did it!

On our first day of field work we found a cliff full of Triassic reptile bones. . . .

This happened yesterday.

The bones are not articulated—it is a stream channel deposit. But they are numerous and in *good* condition. Should give a varied fauna.

We are getting prepared for an intensive collecting program.

The locality is Coalsack Bluff. Only a few miles from our camp. The cliff faces north into the sun, and is thus relatively warm.

We are tremendously excited.

This really pins down continental drift, in my opinion. Antarctica had to be connected [to Africa, Australia and South Amer-
ica] in the Trias.

A very excited old man

Colbert had every right to be excited. The team had found bones
of terrestrial, mammallike reptiles on a continent that was completely surrounded by vast oceans. These animals had been walkers,
not swimmers. There was no way they could have crossed hundreds
of miles of open sea. Therefore the fossils were mute proof that one
or more connections must have existed between Antarctica and other
continents during early Triassic time, some 160 million years ago.
Further collecting focused on a hunt for identifiable evidence of
species, bloodlines, and gene pools. These were the clues from which
important conclusions would be drawn. The geologists knew they
were playing for very high stakes.

Identification of animal migration routes would show how the
continents were joined in early Triassic time. Beyond that there was
the haunting question of our own ancestry. Mammals must have
evolved from mammallike reptiles during this period. New links in
that story would be exceptionally welcome. Oddly enough, success
came quite unexpectedly on the fourth of December as the result of a
randomly directed hammer blow.

It was as if fate had stepped into the game to direct its course.
Dr. Jim Jensen had been working alone and almost mechanically for
nearly five hours searching the high cliffs and ledges for exposed
bone fragments. The cold was intense enough without a chill factor
of minus thirty-five degrees Fahrenheit. Jensen's extremities were
dangerously numb, and his brain dulled and comparatively inactive,
when he surprised himself by impulsively breaking open the apparently homogeneous rock of the cliff face. The newly exposed rock
showed nothing; so he picked up the loose fragment and struck it
another random blow. The rock split open revealing the fossilized
right upper jawbone and curved clawlike tooth of a representative of
a group of mammallike reptiles called the *dicynodonts*. The literal

meaning of this word is "two-dog-tooth." It emphasizes an evolutionary link with mankind, for we also have two canine teeth.

Jensen remembered having found similar fossils at the foot of the Andes Mountains in South America but was too dulled by cold to speculate about it just then. Instead, he reacted matter-of-factly: he wrapped the fossil in toilet paper, marked it for special attention, put it in his collecting bag, and started the long descent back to camp, to warmth, and to food. Hunger, fatigue, and frostbite were the primitive realities of the moment.

Two and a half hours later, Jim Jensen was in the mess hall trying to thaw his painfully frostbitten face when Dr. Colbert, who had seen Jensen's fossil tooth, ran in bursting with excitement! "Jim! This looks like a *lystrosaurus* I found in Africa [and also in India]."

That was enough to excite the roomful of old pros. Up to that time their fieldwork had been a physically demanding hunt. Suddenly it had become an intellectual puzzle as well. The new fossil had added scientific meaning: India and Antarctica were proven now to have been joined as a single continental unit 160 million years earlier. No one had to ask, "Do you really believe that?" The facts spoke for themselves. Because of them, there was no other way to look at the distribution of the *lystrosaurus* gene pool.

Dr. Jensen found himself living three distinct, simultaneous adventures at three distinct levels of consciousness: the agonizing pain of his frostbitten face was a physical reality hard to ignore (it would have been all encompassing had the other sensations not been equally powerful); the thrill of having found the *lystrosaurus* tooth was just being realized (he had done it under conditions that could have driven him off the cliffs: this is the stuff of which heroes are made); the tooth had been brought down from Coalsack Bluff, and because of it whole continents were being moved in the mess hall (an adventure at the intellectual level, where facts are bricks and conclusions are built).

Here we see the link between scientific research and scientific reasoning. Conclusions that are too astonishing to be accepted, yet too interesting to be forgotten, serve to stimulate an ongoing hunt for more facts. This process is beautifully illustrated by Colbert's expedition. At its inception, all he had to go on was the presence of coal as an indicator of a temperate climate and Peter Barrett's word that he had discovered a bone fragment in Antarctica. At its conclusion, Jim Jensen's discovery of the *lystrosaurus* tooth proved that continental displacement is a fact rather than a romantic notion too astonishing to be given further consideration.

Scientific reasoning is a major part of good fieldwork. It represents the culmination of the entire effort. It's the level where conclusions are drawn and learning takes place. *Scientists learn by recognizing the probable truth of each conclusion as they are forced to admit that there is no other more reasonable way to assemble the facts at hand.*

I once asked Dr. Jensen how he felt about this incident. He told me that the thing that has given him the most pleasure was the knowledge that his discovery had encouraged other people to publish their own contributions. Until that time the subject of continental displacement was not taken very seriously by the majority of geologists. Suddenly a fossil tooth in a remote part of the world was moved to center stage of a growing scientific *drama.*

Continental displacement is not so remote an idea as might be imagined. All of us live with the results of it. The quality of life in the have and have-not nations is foreshadowed to a large degree by what has happened to their landmasses. (Chapter 9 shows you how to see some of these things for yourself.)

Geology and Romance

Sometimes a little imagination allows us to enjoy reliving a particularly interesting field trip that someone else has taken. One of my favorites is a "loaf of bread and jug of wine" situation that occurred about 1730 in Somerset in the southwest of England. The heroine was Mrs. Mary Chandler, a lively milliner and poetess who published this astounding verse in 1734.

"Description of Bath"

The shatter'd Rocks and Strata seem to say,
"Nature is old, and tends to her Decay":
Yet, lovely in Decay and green in Age,
Her Beauty lasts to her latest Stage.

The striking thing about this verse is the way that Mary Chandler has combined an accurate description with an awareness of the way rocks indicate the passage of time. When I first read the poem I couldn't imagine how its author was able to learn enough geology to think this way. Who was her teacher?

Several years after discovering Mary's poem, I was probing some broader questions in the history of geology and by accident found what I imagine to be the answer. A modern geologist, John G.C.M. Fuller, published a paper in 1969 with the formidable title "The Industrial Basis of Stratigraphy: John Strachey, 1671–1743, and William Smith, 1769–1839." Fuller's sources included two very old papers written by a coal mining engineer named John Strachey in 1719 and 1725 and published in the *Philosophical Transactions of the Royal Society of London.* Strachey proved to be a geological genius with broad field experience.

Strachey's engineering assignments took him to a number of coal mines in the northern part of Somerset near the beautiful old Roman town of Bath, where Mary lived. The hills in this area are capped by flat-lying rocks that partially cover a set of steeply dipping coal seams visible only in the valleys. Strachey made exhaustive notes on the distribution and attitudes of all these rocks as they were exposed both above ground and in the mine shafts and tunnels. His data were presented in two magnificent vertical cross-sections (see figure 8-2) which form a three-dimensional model of this area. Strachey even tried to explain the origin of the tilted strata by proposing that the beds were wrapped over one another in response to the eastward rotation of the earth. This idea of a two-stage history in which deposition was followed by tilting (i.e., Mary's "The shatter'd

Rocks and Strata . . . ") offers a tenuous thread connecting geologic hero and heroine.

Apparently, no one else in that part of Somerset had this much technical and poetic interest in geology. Therefore it is highly probable that John and Mary knew one another as teacher and student. Could it also have been that the fifty-nine-year-old engineer-geologist and the vivacious forty-three-year-old poetess had a romantic attachment? John Strachey, an ancestor of Lytton Strachey, was a well-known rake in his day. He was a busy man with nineteen children by two wives; perhaps he needed to get away from time to time. If so, would either he or Mary have suspected that they could be caught 250 years after the fact? I have wandered like a musing Omar Khayyam through the lush meadows, shady forests, and quiet lanes of Somerset hoping to be able to tip my hat at their passing carriage. Field trips are great fun!

A good field geologist must learn to practice the art that Shakespeare called "jumping o'er times." It helps establish the validity of rock history. That's how we can learn to escape from the culture-bound time trap into which we were born. Those of us who excel at the art of jumping o'er times establish logical, yet emotionally exciting bonds with both past and future. At the very best it is even possible to imagine living in all times.

It's very difficult for geologists to take vacations from science because the wonder-world is always with us, at times in a most assertive way. Take, for instance, a geological eye-opener in San Francisco at 5:00 in the morning on April 19, 1906. Here is the reaction of Grover Karl Gilbert (1843–1918), a world-class field geologist with a poet's capacity to savor every new opportunity.

> It is the natural and legitimate ambition of a properly constituted geologist to see a glacier, witness an eruption and feel an earthquake. The glacier is always ready, awaiting his visit; the eruption has a course to run, and alacrity only is needed to catch its more important phases; but the earthquake, unheralded and brief, may elude him through his entire lifetime. It had been my fortune

to experience only a single weak tremor, and I had, moreover, been tantalized by narrowly missing the great Inyo [California] earthquake of 1872 and the Alaska earthquake of 1872 and the Alaska earthquake of 1899. When, therefore, I was awakened in Berkeley on the 19th of April last by a tumult of motions and noises, it was with unalloyed pleasure that I became aware that a vigorous earthquake was in progress. The creaking of the building, which has a heavy frame of redwood, and the rattling of various articles of furniture so occupied my attention that I did not fully differentiate the noises peculiar to the earthquake itself. The motions I was able to analyze more successfully, perceiving that, while they had many directions, the dominant factor was a swaying in the north-south direction, which caused me to roll slightly as I lay with my head toward the east. Afterward I found a suspended electric lamp swinging in a north-south direction, and observed that water had been splashed southward from a pitcher.

In my immediate vicinity the destructive effects were trivial, and I did not learn until two hours later that San Francisco was in flames. This information at once incited a tour of observation, and thus began, so far as I was personally concerned, the investigation of the earthquake. A similar beginning was doubtless made by every other geologist in the state, and the initial work of observation and record was individual and without concert. But organization soon followed, and by the end of the second day it is probable that twenty men were working in cooperation under the leadership of Professor J. C. Banner of Stanford University and Professor A. C. Lawson of the State University at Berkeley. . . .

What an exciting day they must have had. The earth's crust had broken along a remarkably straight line, and the two parts had slipped laterally past one another as much as twenty feet. Roads and fences were offset where they crossed the great crack. At first the event was considered to be a local phenomenon centered on San Francisco. Within a few years geologists discovered that the earthquake was simply one local episode on the great San Andreas fault extending from the Pacific coast at Point Reyes, California, southward to the Gulf of California. Seventy years later geologists knew that the movement had been occurring progressively for tens of millions of years with a cumulative displacement of 196 miles in the last 23.5 million years. Eventually they discovered that the San Andreas fault marks the boundary separating that part of the earth's crust containing North America from a separate crustal block that underlies the bulk of the Pacific Ocean. West of the fault, San Francisco and Los Angeles are marching toward Alaska, while Fresno on the east side of the fault stands relatively still. This is continental displacement on a grand scale.

Notice how our own perspectives are changing. A few pages back the talk of proving continental displacement on the basis of a single fossil tooth may have seemed farfetched. All that's changed now. Earthquakes are involved, and we have begun to see that Grove Karl Gilbert in California in 1906 and Edwin H. Colbert in Antarctica in 1969 were studying different parts of the same mechanical system. Unfortunately, Gilbert didn't live to see his work placed in this larger context.

Science is the progressive discovery of the nature of nature. This definition implies that science is open-ended. The people who work at it are hunters, gatherers, and travelers who try to find and understand things they have never known before. That's appealing. Geology is a perfect example of a progressive science. Its ancient Greek roots, *geo* ("earth") and *logy* ("discourse") fix the limits of this adventure to the rocks, minerals, fossils, structures, land forms, oceans, atmosphere, and processes of our planet. All of this was known long ago.

The Reverend John Walker who taught the first systematic course in geology at the University of Edinburgh from 1781 to 1803 told his students exactly what they had to do: "The objects of nature sedulously [i.e., zealously] examined in their native state, the fields and mountains must be traversed, the woods and waters explored, the ocean must be fathomed and its shores scrutinized by everyone that would become proficient in natural knowledge. The way to knowledge of natural history is to go to the fields, the mountains, the oceans, and observe, collect, identify, experiment and study."

That takes a good deal of physical and intellectual energy. John Walker's point was demonstrated to me very clearly years ago when I was serving as an unpaid peon with my first field party. We were working in the Blue Ridge Mountains of Virginia and a famous geologist dropped by to help us. Her name was Anna Jonas Stose, and she was agile as a mountain goat. Our first stop that morning was at an exposure of weathered rock in a deep road cut being made for the then-unfinished Skyline Drive. Anna amazed me by scrambling thirty feet up the embankment to stand on a loose boulder and squint at a piece of rock through her hand lens. Just as she was pronouncing the specimen to be, "Wissahickon Schist . . ." the rock she was standing on began to turn and fall. I shouted a warning but Anna didn't even look up. She just stepped nimbly over to the next foothold and finished her sentence, " . . . there's no doubt about it." I've loved her ever since.

Anna showed me another endearing quality that every hunter must possess. It had been a long, sweaty day, and I had been riding on the runningboard of our field car, holding onto the open doorframe for balance. The windows were down and the dust of the unpaved mountain roads was as bad inside the car as it was out. We all resembled sweat-stained gingerbread cookies. About 4:30 in the afternoon we stopped at a cool spot where the James River had cut a spectacular gorge through the nearly 600 million-year-old Cambrian quartzites. Most of the group began to look for trilobite fossils, but Anna would have none of it.

"John, by this time of day I don't care if I never see another rock. However, I'm quitting for good whenever the excitement doesn't return again by 8:00 A.M. the next day." There we have it: the lure of the hunt is in the mind of the hunter.

Just imagine that you have stumbled on an outcrop of the approximately 375 million-year-old middle Devonian silica formation in northwestern Ohio. A section of the ancient sea floor has been exposed by erosion, and there you see this lovely trilobite of the genus

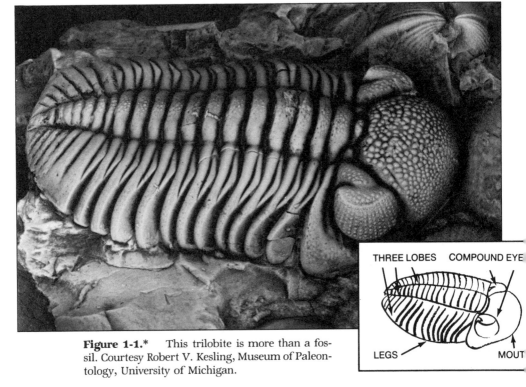

Figure 1-1.* This trilobite is more than a fossil. Courtesy Robert V. Kesling, Museum of Paleontology, University of Michigan.

THREE LOBES   COMPOUND EYE

LEGS   MOUT

Phacops—from the Greek *phakos* ("lentil," "lens") and *ops* ("eye")—staring back at you. This animal is related to modern arthropods such as crabs and insects. It was a go-getter that shared the same body symmetry we have. Would you be tempted to write a better quatrain than Timothy Conrad's (1803–1877) "To a Trilobite"?

> Methinks I see thee gazing from the stone
> With those great eyes, and smiling in scorn
> Of notions and of systems which have grown
> From relics of the time when thou wert born.

If so, you have grasped the message of this chapter. Geologists are hunters of *things* in order to be hunters of the *mind*.

*All sketches show what to look for

TWO

Dispelling the Mystique of Scientific Thinking

"There's nothing worse than thinking you're thinking and not knowing what you're thinking about."

MISTA BOB CATES

C hildren learn to think for themselves through a combination of unabashed curiosity and a confident realization that learning is possible. They are rarely discouraged as long as the pressure to learn is self-imposed. School learning is something else again. My own discovery of this came by accident when I was six years old and in the first grade.

My first-grade class was full of highly animated, very bright youngsters. Any hope that my parents might have had that their son would occupy the scholar's seat—number one on the first row—was totally unrealistic. The competition was too stiff. However, for a short time even I harbored ambitions of academic recognition. Ravaging slash marks in red ink and a grade of thirty on my first spelling test ended that dream. I recognized that my scholastic career was ruined and turned my thoughts in more practical directions. There was no point in taking this spelling test home and becoming a living-room battle casualty.

33

Individual survival is a gene-driven force, and I responded to it with determination: "bury the evidence." As I knelt in a lonely spot to commit this crime against integrity, I saw the most beautiful rock imaginable. It can only be described in purple prose: my rock was tinted by the setting sun a glistening mass of golden, glassy grains that sparkled and flashed in a wondrous way. It made me think of a picture I had seen of the Hall of Mirrors in Versailles. The spelling test was forgotten as a new fantasy world spread before me. What was this thing?

I took it home to my father, an electrical engineer, and at that time the most all-knowing Moses in my life. Pop hefted my rock with all of the understanding of a man kicking an automobile tire, squinted a little as if welcoming an old friend, and pronounced it "a piece of dried mud."

Even a first-grader knew better. Mud is a dirty mixture with a lot of clay particles in it. My rock wasn't dirty. It glistened in the lamplight and was so transparent I could see through the thin broken edges. Perhaps I was doomed to be a hopeless student on the beaten paths of schoolwork, but I knew one thing: my father was wrong! He knew even less than I did because he wasn't aware that he *was* wrong. This thought gave me great pleasure.

That was fifty-eight years ago, yet I still remember how thrilled I was to be superior to him in at least one respect. It marks the moment that I began to trust my own judgment in matters of what does and what does not make sense. With that single triumph I began to learn to dispel the mystique of what may be called scientific thinking. All good thinking must be based on common sense and not on the ex cathedra pronouncements of authority figures.

Sir Charles Lyell (1797–1875), the most respected British geologist of the nineteenth century, summed all of this up in two maxims that guided his professional life: "go and see" and "prefer reason to authority."

A few months later while digging around in a partially completed foundation hole, I found another piece of rock, much like the first but weathered enough to be pounded to crumbs between two bricks. I was delighted to discover that the grains could be separated from one another, becoming once again a handful of loose sand. Sir Charles would have been proud of me.

Here was the real answer to the question that had defeated my father. My beautiful, grainy, glassy rock was sand that somehow had been welded together. Pounding may not have been a very scientific method, but it had led me to a new level of understanding. I had made a great discovery: rocks have histories!

Imagine what an appreciative teacher might have been able to do for me at that point. I had reached a stage where I had begun to transfer a single observation into a broad general context. Unfortunately, no one was around to give me that kind of direction, to say, "Hey, kid, that's good. Look what you've just done. Keep it up!" A teacher could have pointed out that my rock was called quartzite and was characterized by the way it broke *across* the sand grains rather than *around* them. This would have introduced me to the idea of systematic classification in science and the value of words and definitions to help me pull similar things out of books so that I might compare them. Teachers open doors and show us how to enter the real world that lies just beyond our knowledge. They do this for the humblest and greatest of scholars.

Linus Pauling, winner of two Nobel prizes (one in chemistry, 1954, and the Peace Prize, 1962), recalls:

> When I was about six years old we lived in Condon, a small town in eastern Oregon. One day I was trying to sharpen a pencil with my knife. I was unsuccessful until a cowboy said that he would show me how. He pointed out that there was a relationship between the angle at which I held the blade of the knife and the slice of wood that was removed from the pencil. I was so impressed by the fact that one could attack a problem in a logical way that I have remembered this episode all my life.

Another anecdote (to be found in the work of Voltaire) places Sir Isaac Newton (1642–1727) in the teacher's role, explaining to the rest of us how to develop our own qualities of genius! "A visitor once asked Newton how he discovered the laws of the world. 'By thinking about it ceaselessly,' he answered. 'This is the secret of all great discoveries: genius in the sciences depends only on the intensity and duration of thought that a man can muster.' "

These stories bring the behavior of two remarkable geniuses into a range that can be understood. The key rests with an elegant definition of the word *understanding*. Typical dictionary definitions are of little help because they are somewhat circular: understanding is said to mean comprehension and comprehension is equated with understanding. A better definition for our purposes would show that understanding is a sport of participation and therefore something of a game in which players such as Newton and Pauling have excelled. The game has only one rule: *draw the least astonishing conclusion that can be supported by the known set of facts.* Think about this rule as you read our definition: *understanding is knowing enough about something so that each part of it is seen in proper relationship to every other part and to the whole thing.*

My role at the moment is to serve as a teacher and to show exactly how supporting facts (or parts) are assembled to form conclusions (or the whole thing). If I can do that properly, much of the mystique of scientific thinking will disappear, and we will be free to see the world as geologists do. Sherlock Holmes will help us. His reasoning was exactly the same as that of Newton and Pauling: straightforward and absolutely devoid of mystique. This classic example is taken from Arthur Conan Doyle's story, "A Study in Scarlet." This was Holmes and Watson's first adventure together. Dr. Watson had not known Holmes long and was doubtful that the man really had the genius to see where others were blind.

The scene opens at 221B Baker Street, London. Holmes and Watson are looking out the window. Watson is speaking.

> "I wonder what that fellow is looking for?" I asked, pointing
> to a stalwart, plainly dressed individual who was walking slowly
> down the other side of the street, looking anxiously at the num-

bers. He had a large blue envelope in his hand, and was evidently the bearer of a message.

"You mean that retired sergeant of Marines," said Sherlock Holmes.

"Brag and bounce!" thought I to myself. "He knows that I cannot verify his guess."

The thought had hardly passed through my mind when the man whom we were watching caught sight of the number on our door and ran rapidly across the roadway. We heard a loud knock, a deep voice below, and heavy steps ascending the stair.

"For Mr. Sherlock Holmes," he said, stepping into the room and handing my friend the letter.

Here was an opportunity for taking the conceit out of him. He little thought of this when he made that random shot.

"May I ask, my lad," I said blandly, "what your trade may be?"

"Commissionnaire, Sir," he said gruffly. "Uniform away for repairs."

"And you were?" I asked with a slightly malicious glance at my companion.

"Sergeant, sir, Royal Marine Light Infantry, sir. No answer? Right, sir." He clicked his heels together, raised his hand in salute, and was gone. . . .

"How in the world did you deduce that?" I asked. . . .

"It was easier to know it than to explain why I know it. If you were asked to prove that two and two made four, you might have some difficulty, and yet be quite sure of the fact. Even across the street I could see a great, blue anchor tatooed on the back of the fellow's hand. That smacked of the sea. He had a military carriage, however, and the regulation sidewhiskers. There we have the marine. He was a man of some amount of self-importance and a certain air of command. You must have observed the way in which he held his head and swung his cane. A steady, respectable middle-aged man, too, on the face of him—all facts which led me to believe that he had been a sergeant."

Sherlock Holmes selected a set of facts, such as the great blue tatoo and military carriage, and from them drew a conclusion based on common experience. His method was to think of a single context that would be the obvious sum of all the known parts. Holmes recognized a sergeant of marines as the least astonishing choice. Solicitor, professional wine taster, and Archbishop of Canterbury were possible, but totally improbable choices when compared with the simple characteristics of sergeants of marines. Every least astonishing conclusion is a winner, judged to be the most probable choice of all available competitors. Astounding as it may seem, that is the limit of precision to be found among all the sciences, even those that have been called exact!

Part of the charm of the Sherlock Holmes stories lies in the way the author, Sir Arthur Conan Doyle, startled his readers, first, by having Holmes state his deductions, and, then, by showing how simply he explained them. Science writers can use the same pattern with equal success if readers understand this simple relationship:

$$\text{Facts } 1+2+3+4+\ldots+n \to \begin{bmatrix} \text{when considered together} \\ \text{imply least astonishing} \\ \text{conclusion} \end{bmatrix} \to \begin{bmatrix} \text{specific conclusion(s)} \\ \text{stated as model(s)} \\ \text{of nature} \end{bmatrix}$$

This system of analysis is very old. Aesop used it in his fables twenty-five hundred years ago. His story of the blind men and the elephant is typical. Each of the blind men felt a different part of the elephant and concluded that the whole beast resembled his one portion. There must be some coordination of experience before the correct conclusion (i.e., the whole elephant) can be deduced from the parts through the principle of least astonishment. Lacking this coordination, the blind men postulated a wide range of untenable, hypothetical models.

The system that works so well in fiction is almost identical to the one that scientists use. The chief difference is that scientists attempt to define their facts in absolute terms and recognize least astonishing solutions against a background of shared technical, rather than common human, experience. The use of technical experience that is often beyond the understanding of laypersons has given science a reputation for practicing mumbo jumbo. The only way to eliminate this confusion is to educate laity to think as scientists do. There are three parts in this system of scientific thinking: 1) finding the facts; 2) selecting the least astonishing explanation(s); and 3) deriving a model of nature.

Facts are the basic building blocks of science. We need to understand what they are. A philosopher might tell us that a fact is a reference point of experience. An even better definition was given to me by Sergeant Sanders, a Pinkerton guard I met in an art gallery. He had a way of getting right to the heart of a question. "A fact is a sure thing. You can bet on it." That's clear enough.

Most of the mystique to be dispelled in scientific thinking rests with the vagueness of selecting one explanation over another to form a model of nature. That difficulty is real. It bothers professional scientists as much as it may bother concerned laypersons. However, the professionals are very polite about the problem and have established a private language to deal with it. When two groups of scientists disagree about an interpretation of a set of facts, they call it a controversy. This means that what seems to be least astonishing to one group is rejected as far too astonishing by the other. Solutions to this standoff don't rest in majority opinion, debate, or oratory. Both groups should realize what they need are more facts and a shared perspective.

The battle being waged between evolutionists and creationists is a marvelous case in point. We can expect it to last for decades, if not centuries. At present the two groups are talking past each other. Creationists assert divine intervention in the creation of mankind. Devoted evolutionists deny the necessity of divine intervention at any stage of development. There is no way for these groups to agree and at the same time remain apart on this single value judgment of what is or is not least astonishing.

Leonardo da Vinci (1452–1519) faced and dismissed a similar problem when he struggled with the illogical character of Noah's supposedly worldwide flood. The two points that concerned him most were the observed distribution of marine fossils far inland and the possible (or impossible) destination of the receding water. He approached these problems methodically.

Clam fossils were known to be present in rocks that in Leonardo's day lay far from the sea. He experimented with living clams and found that they could move about twelve feet a day: "Therefore, with such a rate of motion it would not have traveled from the Adriatic Sea as far as Monferrato in Lombardy, a distance of two hundred and fifty miles, in forty days—as he said who kept a record of this time." With that, he dismissed the accuracy of the account.

His second task was to calculate the amount of water necessary for a flood of Biblical magnitude. He recognized that water flooding a spherical earth to a depth of fifteen feet above the highest mountain would have nowhere to run off in the way that a flooded river valley is drained into the sea. "How then did the waters of so great a Flood depart if it is proved that they had no power of motion? If it departed, how did it move, unless it went upwards? At this point natural causes fail us, and therefore in order to resolve such a doubt we must needs either call in a miracle to our aid or else say that all of this water was evaporated by the heat of the sun." This entry in Leonardo's notebooks is listed under the heading "Doubt." Obviously, miracles were not part of his conception of science.

Questions of ambiguity are difficult to deal with. Scientists have set up a very simple classification system to identify ambiguities and keep them at arm's length. The spectrum extends from the certainties of scientific laws at one end to vague hypotheses at the other. Major models of nature, accepted by everyone because there are no known exceptions, are called laws. Most laws are restricted to the sciences of chemistry and physics because their experimental methods are capable of eliminating ambiguities far better than the research methods of the other sciences. Theories represent a much lower level of certainty. A good theory, such as organic evolution, is supported by a vast body of facts, yet questions remain unresolved.

There is little point in debating the validity of a theory except to point out the additional facts that are needed to prove or disprove it.

Hypotheses are even more vaguely supported models of nature. Once again the need is to establish more facts as a means either of destroying the hypothesis or of elevating its rank to that of a theory. No wonder the definition of science includes the idea of progressive discovery. The business of science is research because science is an unfinished business.

Much of the mystique of scientific thinking disappears as soon as we realize that no claims of omnipotence are made for it. There is no need to apologize for fallibility as long as levels of ignorance are properly acknowledged. There are very few laws and a tremendous number of wishy-washy hypotheses. Controversies abound. Ignorance is rife. Science is a social subculture that operates very effectively because scientists share the same experiences and the same ways of doing things. Scientists think of themselves as hunters struggling in poorly known jungles rather than pontificators on pedestals of past successes. The doctrine of science is that there must be no doctrine, no creed, and no constricting oath beginning with the words "I believe. . . ." Creative scientists work in an open-ended world of splendid uncertainty. Geologists are sometimes pitied, for their world is the most uncertain of all. Theirs is a historical science, rooted in a past that defies experimental research methods.

At one end of the time spectrum geologists work on problems as contemporary as the eruption of Mount Saint Helens in Washington State (see chapter 9). Yet they must also work with the processes of mountain building to which this episode belongs. That series of events began tens of millions of years ago as two great plates of the earth's crust began to grind against each other. Present and past are linked in a continuum of events. Geologists cannot use the experimental methods of physics and chemistry; continents and eons fit neither the test tube nor a life span. As a result, geological research is reduced to reconstructing the past from fragmentary information and the best dates attainable.

Geologists are forced to think backward from the present to the past on the basis of incomplete data, a task comparable to playing gin rummy with a deck of fifty-one (or fewer) cards. Some of the rock record is concealed, an enormous part, destroyed. Only the most intellectually daring scientists could hope to be able to challenge Rudyard Kipling's cynicism:

> Ah! What avails the classic bent
>   And what the cultured word,
> Against the undoctored incident
>   That actually occurred?

Nevertheless we try.

Some of the Neo-Confucian Chinese scholars of the twelfth century A.D. were very skilled in the practice of thinking backward, as illustrated by this example from the writings of Shen Kua.

> In recent years [circa A.D. 1180] there was a landslide of the bank of a large river in Yung-Ning Kuam near Yenchow. The bank collapsed, opening a space of several dozens of feet, and under the ground a forest of bamboo shoots was revealed. It contained several hundred bamboos with their roots and trunks all complete and all turned to stone. A high official happened to pass by and took away several saying that he would present them to the Emperor. Now bamboos do not grow in Yenchow. These were several dozens of feet below the present surface of the ground, and we do not know in what dynasty they could possibly have grown. Perhaps, in very ancient times the climate was different so that the place was low, damp, gloomy and suitable for bamboos.

Shen Kua produced a perfectly reasonable, least-astonishing deduction about the paleoclimate in Yenchow. The most remarkable thing about his thinking, however, was that he felt the facts demanded an explanation (most people would do little more than dismiss the stone bamboos as another wonder story), and his approach seems to have been as systematic as that of any modern scientist. First, he recognized that the shapes of the stone bamboos and living bamboos were identical and therefore constituted a problem demanding explanation. By declaring that both were bamboos, in effect, he brought stone to life. Having done that, he was forced to put them into a time frame. Present time and future time were both impossible; so the fossils were construed to have lived in the past. It was at that point that Shen Kua must have had his second remarkable insight! "Bamboos do not grow in Yenchow." A second problem; a second judgment required: "Perhaps, in very ancient times, the climate was different so that the place was low, damp, gloomy and suitable for bamboos." The paths of genius are beautiful things to trace!

Nicolaus Steno (1638–1687), a Danish anatomist serving with the court of Ferdinand II of Tuscany, made exactly the same sort of intellectual leap. He had been given the head of a great white shark to dissect and describe. He found its teeth to be particularly interesting for they resembled strange stone objects that had been discovered in the rocks of the Maltese Islands. Once again, the linkage from living teeth to stone ones was made by way of appearance and a judgment that brought the fossils to life in ancient times. We will see a good deal more of Steno, for it was he who gave us a number of techniques to decipher the history of rocks.

Scottish Roots of Geologic Understanding

Three splendid Scots—James Hutton, John Playfair, and Sir Charles Lyell—working in the late eighteenth and early nineteenth centuries added to Steno's insights and established a historical viewpoint we now call the principle of uniformity. This is a method of reasoning by analogy from the present to the past. It's nothing more than developing a least-astonishing judgment that geologic processes actively producing observable results today must have operated in the past as well wherever we find identical structures of both ages.

The principle of uniformity was first formalized by Hutton (1726–1797) as the result of various field excursions made in the company of gifted friends. His conclusions were published in 1785 in a revolutionary book entitled *Theory of the Earth.* The last line of the book is among the finest in scientific literature and offers a remarkable summary of his contributions to geology:

> The result, therefore, of our present enquiry is, that we find
> no vestige of a beginning, no prospect of an end.

That line will mean more to us after we have been introduced to Charles Darwin's (1809–1882) revolutionary ideas published in 1859 as *The Origin of Species.* Darwin's theory presupposed a vast time in which organic evolution could take place. Hutton gave it to him. Science operates in this way. One breakthrough leads to another. One of my teachers called the sequence of events "steamboat time": Robert Fulton could not have invented a practical steamboat until James Watt had invented a practical steam engine and other predecessors had invented machine tools to fashion the steel parts. Genius both feeds and builds on genius.

A more complete and readable view of the earth as an operating machine was published in 1802 by John Playfair (1748–1819), a close friend of Hutton's. His book, entitled *Illustrations of the Huttonian Theory of the Earth,* contains detailed observations that were not properly explained in the original text.

Charles Lyell was enthralled by the principle of uniformity. Here was a tool that allowed him to use his own analytical abilities, so well expressed in the dictum "prefer reason to authority." By the time Lyell was thirty in 1827, he had determined that the course of his life required him to write a book on the principles of geology: "I am going to write in confirmation of ancient causes having been the same as modern." He spent the next year or so traveling across Europe and down the boot of Italy toward the volcanoes of Sicily. In one letter to his father, he wrote: "The whole tour has been rich, as I had anticipated . . . in those analogies between existing nature and the effects of causes in remote areas which it will be the great object of my work to point out. . . ."

The young man was on his way to creating the intellectual revolution that Hutton and Playfair had shown to be a possibility. Lyell's contribution, *Principles of Geology,* was published in three volumes between 1830 and 1833. The essence of the book is illustrated in a woodcut of the ruins of the pre-Christian Temple of Serapis at Pozzuoli, a few miles west of Naples.

The dark markings about nineteen feet about floor level were made by a boring marine mollusk of the genus *Lithodomus*—from the Greek *lithos* ("stone") and the Latin *domus* ("house"). Many shells of these creatures are still to be found embedded in the marble of the pillars. These facts, like Shen Kua's bamboos, can be explained quite easily by the principle of uniformity.

The history of the temple began a few centuries before the Christian era, when it was built by the Romans on land that was well above sea level. The temple was then submerged beneath the

Figure 2-1. The Temple of Serapis, Pozzuoli, Italy, as Charles Lyell saw it in 1828, *left,* and as it appears today, *right.*

sea by a localized twenty-five-foot lowering of the land. At this time the marine mollusks attached themselves to the pillars to a depth of seven or eight feet below sea level and bored their way into the marble. Later, the sea floor was lifted to about its original elevation and the temple was exposed. Charles Lyell saw this as a beautiful illustration of the vertical instability of the earth. From there it was an easy step to show that sedimentary rocks containing fossil shells of marine animals could also be explained by vertical movements of the earth's crust.

I visited the temple in October 1981 and was astounded to see how chemical weathering has damaged the pillars since the 1830s. The explanation serves as an additional example of the use of the principle of uniformity. Lyell was not able to imagine the atmospheric changes that were to take place as a result of the industrial revolution, but apparently chemicals added to the air over the throbbing city of Naples have attacked the pillars mercilessly. One pillar is now little more than a ruin enclosed in an iron cage. The temple should be seen in four distinct time frames: the time of construction, the time of sinking land and simultaneous marine invasion, the time of uplift of the land, and, finally, the time of industrial air pollution.

If you can see these distinctions, you have learned to look at things as Shen Kua did. It's the level of intellectual alertness that must be reached before the earth becomes a puzzle waiting to be solved. The next level of intellectual alertness is reached when we start creating hypothetical explanations to serve as models of the way that nature should be. This is the beginning of creative science. At this level we become sensitive to new facts that will either strengthen or destroy our hypotheses.

Research is analogous to looking for a four-leaf clover. Once we know what we are looking for, we have a greater chance of recognizing it when we see it for the first time. This is the way James Hutton did his field work. It's an interesting story, particularly when seen in context.

In Hutton's day Abraham Gottlon Werner (1749–1817), a professor in a German mining academy, was the absolute authority on the origin of crystalline rocks. Werner thought that massive granites were the most ancient of all rocks, that they had been precipitated as sediments on the floor of a once-worldwide ocean. In his model of the earth, massive granites formed the basement platform on which all younger rocks had been deposited in successively younger layers. Hutton was suspicious of all of this, because he had found veins of granite cutting across older, layered rock structures in the beautiful northwestern Highlands of Scotland beyond the Great Glen (Glenmore). Such a thing couldn't happen in the rock order imagined by Professor Werner. Hutton felt that granites may have formed from molten masses of rock that were once fluid enough to have been injected as liquids in preexisting cracks in the surrounding rocks. A controversy was building, and the stakes were high.

Before we become too technical, we need to pause for a moment to determine what was taking place in Hutton's mind. The story is less [one] of rocks than of imagination. The northwestern Highlands are a rugged region of deep valleys, rushing streams, high crags, and heather-covered slopes extending far above timberline. It's a soggy place. There is water everywhere—in the lakes, in the rivulets. Water trickles from cracks in rocks and squeezes out beneath your feet with every step. Clouds, in slate-colored layers and white, two-story masses, are constantly on the move, casting shadows and releasing driving rain. All of this is interspersed with brilliant patches of sunlight. The scenery of Scotland must be seen to be believed. It's incredible. Stand on some high point and turn the full 360 degrees. The entire mixture is overwhelmingly beautiful.

Put Hutton in the center of this scene (see portrait, page 15) with his long nose pressed against the details of some never-before-visited rock outcrop; the cold, the damp, even the breathtaking vistas, ignored. Hutton's imagination transcends both time and space, transforms a vein of cold, hard granite into an exceedingly hot, liquid mush at least twelve miles below the surface of the earth. This, to him, was the least astonishing judgment.

In 1785 James Hutton was embarked on the opening phase of a definitive study of igneous rocks, which continues to this day. (A point illustrated in chapter 9 with the discussion of the eruption of Mount Saint Helens in Washington.) He was determined to eliminate controversy with Werner by supporting or denying an igneous—from the Latin *igneus* ("five")—origin for granite. (Liquid granite suggests lava and volcanoes.) He knew that the first step was to find better exposures and more dependable field evidence.

> Having thus suspended my opinion [Hutton wrote] until I should have an opportunity of finding some decisive appearance [that is, an outcrop], by which this important question might be determined with certainty, I considered where it might be most likely to find the junction of the granite country with the Alpine strata [that is, folded sedimentary rocks]. Mr. Clerk, of Eldin, and I had an engagement to visit the Duke of Atholl, at Blair. I concluded that from Blair [in the Grampian Mountains of Scotland] it could not be far before the great mass of granite, which runs south-west from Aberdeen, would be met with, in ascending the river Tilt, or some of its branches. Mr. Clerk and I were, however, resolved to find it out, to whatever distance the pursuit might lead us among the mountains of this elevated track. Little did we imagine that we should be so fortunate as to meet with the object of our search almost upon the very spot where the Duke's hunting-seat is situate, and where we were entertained with the utmost hospitality and elegance.
>
> It is in Glen Tilt, and precisely in the bed of the river, that this junction is formed of the granite with the Alpine strata. . . . I here had every satisfaction that it was possible to desire, having found the most perfect evidence, that granite had been made to break the Alpine strata, and invade the country in a fluid state.

Hutton's writings are easily overlooked because his style often lacks descriptive precision. He told us what he saw, and yet we're still not exactly sure how it looked. Imprecision is an odd quality to find in a man who still holds his position as a critical thinker after the passage of two centuries. There is a reason however: Hutton's work was intended to be illustrated profusely with very accurate line drawings and watercolors made by his colleague and field assistant, the eminent John Clerk of Eldin Castle (1728–1812). Clerk owned coal mines and was involved in other engineering projects that must have sharpened his ability to function as an observer of unusual ability. Clerk's drawings were intended to fill the gaps in Hutton's prose. Unfortunately, something went astray. The illustrations were lost and Hutton's publications were printed without them. No one even suspected the drawings still existed until August 1968, when Sir John Clerk of Penicuick House, Midlothian, found seventy of them in an old folio among the family papers. The geological community was alerted and James Hutton was reborn.

Charles Waterston of the Royal Scottish Museum, Edinburgh, passed the folio on to Gordon Craig, Hutton Professor of Geology, University of Edinburgh, and Donald B. McIntyre, professor of geology, Pomona College, Claremont, California, to be edited and published at long last. Facsimiles are now available under the title *James Hutton's Theory of the Earth: The Lost Drawings.* All of this discovering, editing, writing, and publishing was going on in Edinburgh in 1976 without my knowledge.

At the time, I was working (in what proved to be intellectual isolation) on a textbook designed to show students where geological knowledge came from. Hutton's genius was still a mystery to me: so I took a trip to Glen Tilt to see the rocks for myself. I found the duke of Atholl's hunting lodge located in a private game preserve of about 231,000 acres (631 square miles). Access to the critical outcrop is provided by a narrow gravel road that parallels the River Tilt upstream from the small town of Blair Atholl. The first few miles of the valley are densely wooded with outcrops of tilted sediments visible along the banks. Granite boulders among the river gravels attest to a major source farther upstream. Hutton probably knew this much before he began his pilgrimage. We may sense his excitement at this stage by the fact that the lost drawings contain three pictures of rounded stream pebbles with veins of granite cutting through the bedding planes of older sediments. Again, it was a case of the bleat-

Figure 2-2. Glen Tilt, Scotland, the scene of our linked adventures in 1785 and 1977. Look ahead to figure 3-1 to see how this area appeared toward the end of glacial times.

ing lamb exciting the tiger. My own feeling, as I followed Hutton upstream, was the thrill of being in hot pursuit.

I hadn't gone very far before the valley opened out, first, into a weed-choked meadow with few trees and, then, quite sharply into the great, grassy, *U*-shaped valley form shown in figure 2-2. The sediments along the banks of the river are primarily white quartz-ites, and the granite boulders in the riverbed looked larger and more prominent. After the road crossed the river for the third time, I ran into a group of British botanists scurrying across a footbridge to climb the northwest-facing valley wall. When I asked them what they were doing, an elderly man replied, "Alpine plants have been reported here on the shady and cold side of the valley. We're going to look for them. The competition will be very keen." Suddenly I began to see James Hutton, the man, in a better light. Everything is a game with these people. They participate because it's fun to win.

This is what Hutton was doing in Glen Tilt: competing to win! I pressed on after him, keeping one eye on the valley walls and the other on the riverbed. Suddenly, there it was: orange-red granite exposed in the bed of the river, just as Hutton had said. Excitedly I spent the next hour or so climbing over the rocks and photographing places where veins of granite had intruded the sedimentary layers, breaking them apart and surrounding them with the once-liquid rock. The continuity, from bed to bed, was clearly shown, even though the parts were separated by a foot or more of granite. This was a splendid sight. Mr. McGregor, the duke of Atholl's gilly, came by to share the fun. He told me that I was in the wrong place. All the other geologists who came to Glen Tilt went up-stream two more miles to a washed-out stone bridge beyond the old hunting lodge. I told him, "No, that's impossible. Hutton would never pass this outcrop to find another one." A year later I began to have doubts.

News of the lost drawings finally had filtered down to me; so I determined to go back to Scotland and take another look. I came across the Irish Sea from Larne to Stranraer and spent the night in Robert Burns's hometown of Ayr. The next morning I called the University of Edinburgh and asked the switchboard operator to connect me with the Hutton specialist in the geology department. She put me through to Gordon Craig.

Such is the bond of the geological fraternity that I was invited to his office on the spot. We spent a happy hour together marveling at the details shown in John Clerk's drawings. There was no question that Hutton and Clerk had first seen granite intruding sediments at

Figure 2-3. James Hutton and John Clerk stood here on the now-eroded stone bridge and saw proof that melted rock had intruded these sediments.

an outcrop about a mile upstream from the duke's hunting lodge and about two miles above the outcrop I had visited a year earlier. We came to the conclusion that on their first trip up the glen, Hutton and Clerk must have been overtaken by darkness and ridden their horses on to the hunting lodge without seeing the lower granite mass.

The higher outcrop is a thing of great beauty as shown in fig-

Figure 2-4. These are typical veins of granite cutting across and surrounding sedimentary beds.

ures 2-3, 2-4, and 2-5. John Clerk must have stood on the bridge and sketched the outline of the streambed and then filled in details with measurements made on the rocks. His original drawing is a dramatic watercolor in shades of gray and terra-cotta, the principal colors of the real outcrop. The bridge abutments are shown as rectangular blanks. Notice how well Clerk has shown the fractured and displaced pattern of sediments with the once-liquid granite intruding them. No wonder Hutton felt every satisfaction that it was possible to desire. Billions of people had preceded him on earth, yet no one else had ever proven that granite had once been a hot liquid. A revolution in scientific thought was born that day in 1785. James Hutton had forged the first liquid link between granites formed in the depths of the earth's crust and lavas and volcanoes formed on the surface. Hutton had won the game and, in a sense, we were there to see him do it . . . *sans mystique.*

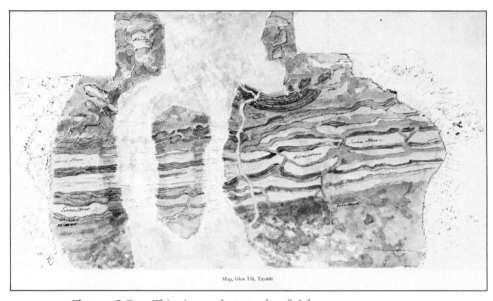

Figure 2-5. This is a photograph of John Clerk's 1785 drawing of granite-invaded sediments at the site of the stone bridge over the Tilt River about a mile upstream from the hunting lodge.

THREE

Using the Principle of Least Astonishment

"There should be another book in the
Bible . . . the Book of How."

OLD SOUTHERN SAYING

A s Sherlock Holmes's powers of deductive reasoning helped us understand the notion of the least astonishing conclusion, the inductive reasoning of another literary figure, Robinson Crusoe, will help us understand the dynamics of intellectual revolutions.

It happened one day, about noon, going toward my boat, I was exceedingly surprised with the print of a man's naked foot on the shore, which was very plain to be seen on the sand. I stood like one thunderstruck, or as if I had seen an apparition.

Having lived in isolation more than twenty years on a very small island, Crusoe had every right to believe that he knew all about his home. For someone else to be on his island was incomprehensible. That footprint was what we call, in scientific terms, an anomaly, a deviation from the expected state of nature. Crusoe's first reaction was one of unrestrained panic. He ran back to his fortress-house and cowered there for several days, fearful that cannibals might take a warm interest in him. Eventually another thought occurred to him: perhaps this was his own long forgotten footprint made on some earlier occasion. Verification was simple enough: return and compare the stray footprint with a new one of his own. That was the scientific thing to do: make a field check.

> I listened, I looked around me, I could hear nothing, nor see anything. . . . I could see no other impression but that one. I went to it again . . . to observe if it might not be my fancy; but there was no room for that, for there was exactly the very print of a foot—toes, heel and every part of a foot . . . it appeared . . . to me when I came to measure the mark with my own foot, I found my foot not so large by a great deal.

After that, there was only one possible explanation: "I went home again, filled with the belief that some man or men had been on the shore there, or, in short, that the island was inhabited."

Robinson Crusoe was a practical man, who paid attention to the anomalous footprint because it was too important to be overlooked. The anomaly represented "an irrepressible crisis of contradiction" that destroyed his old viewpoint and dictated the form of an alternative. We're now ready to place his story into a philosophical content and draw an important generalization from it. Once a crisis of contradiction has been verified, the old viewpoint, or model of the state of nature, must be abandoned; the new viewpoint must include all the facts within a single context. Nothing remains of the old anomaly because it's included now in the new model and acts as a supporting fact with the same rank as all the other facts.

In his classic study, *The Structure of Scientific Revolutions*, Thomas S. Kuhn recognizes that most scientists are engaged in problem solving using techniques and concepts that have already been developed. He calls this work "normal science" and defines it as "research firmly based upon one or more past scientific achievements . . . that some particular scientific community acknowledges for a time as supplying the foundation for its further practice. . . ." Normal science is routine rather than revolutionary because it contains no element of surprise. All judgments applied in it are within the limits of our expectations of the way nature should be. Normal science ends, however, with the verification of an irrepressible crisis of contradiction like Crusoe's.

Kuhn points out that the people who have the imagination to create revolutionary ideas are almost always either young or new to the scientific specialty in which they make their breakthroughs. He attributes this to their lack of commitment to the traditional rules of normal science in these areas. The young and uncommitted are able more freely to discard old ideas and create new ones to take their place. Of course, anyone with sufficient imagination is able to heal the wound left after an old idea has been rooted out. All that is necessary is to create a new, defensible model of nature that explains every fact and leaves no anomalies.

We're going to examine the work of three "Kuhnians"—Louis Agassiz, age thirty, Charles Darwin, age twenty-eight, and Alfred Wegener, age thirty-five—who discovered the Ice Age, the dynamics of organic evolution, and the principle of continental drift, respectively. Their discoveries are examples of the combining of reason with imagination.

Louis Agassiz

Jean Louis Rodolphe Agassiz (1807–1873) had two gifts that characterize most truly great people: high intelligence and unbounded animal energy. The combination is unbeatable in a sensitive person of high moral character. Agassiz was born in a small Swiss village near Neuchatel. By the time he was twenty-three, he had studied in three universities: Zurich, Heidelberg, and Munich; had earned doctorates in botany and medicine; had done the fieldwork and published a book on the freshwater fish of Brazil; and had issued a preliminary statement on a *History of Freshwater Fishes of Central Europe.*

By the time he was twenty-five, Agassiz had been appointed professor of natural history at the University of Neuchatel and was engaged in a major, five-volume study of fossil fish. At twenty-nine, he was presented the highly prized Wollaston Medal by the Geological Society of London. At thirty-one, Agassiz was invited to become a foreign member of the British Royal Society in recognition of his being the premier scholar in the highly specialized study of living and fossil fish. Louis had arrived; yet in the characteristic pattern of many high-energy people, his career had already turned in a new direction in response to a challenging opportunity.

In July 1834, when Louis was twenty-seven, he was present at a meeting of the Swiss Society of Natural Science in Lucerne. An old friend, Jean de Charpentier, read a paper on the mechanics of Alpine glaciation. This was a systematic presentation of things that had been known for several decades to some geologists as well as to many of the local people who lived in the high Alpine valleys. De Charpentier made a case that accounted for a colder climate in the recent past on the basis of a remarkable set of anomalies.

With contemporary photographs from different parts of the world, we can demonstrate that de Charpentier's points were of more than local significance.

Mountain glaciers are literally rivers of ice that flow slowly downhill to melt at some level where the climate is warm enough to destroy them. Figure 3-1 shows that this is true, even in Antarctica. Extensive, polished, and scratched rock surfaces, such as the one in Ireland illustrated in figure 3-2, are found far beyond the present ice limits. Scratches of this sort parallel the axis of the valley and, judging from examples found at the ice front, are made by rock fragments pushed along at the base of the moving ice. Linear banks of jumbled boulders usually occur below each ice front and appear to be extensions of rocks weathered from the valley walls and originally deposited on the moving ice. An example from the Himalayas of Nepal is shown in figure 3-3. Huge, heavy boulders made of rock

Figure 3-1. This is a typical mountain glacier near the coast of Antarctica. Note the deep *U*-shaped valley and the bare ground at the lower end where the melting rate exceeds the rate of snow accumulation and glacial flow.

types that contrast with the local materials, similar to the example shown in figure 3-4, are also found far down the valleys and high above their bottoms. No one could imagine that water had been the agent that transported such big rocks from distant sources to their present localities. Lake basins carved from hard rocks and strung like beads along the strange glacially carved *U*-shaped valley bottoms offered an additional intriguing problem. The one pictured in figure 3-5 is the upper lake at Killarney, County Kerry, Ireland.

Jean de Charpentier saw that these lake basins were particularly important. The hollows couldn't have been cut by rapidly flowing water without the preposterous requirement that water would have to flow uphill at high speed over the hard rock lips at the lower end of each lake. Ponded water is the site of deposition, not erosion. He was convinced that the entire set of anomalies presented an irrepressible crisis of contradiction demanding a new explanation.

The only explanation that made sense to him was one of much more extensive glacial erosion, transportation, and deposition under

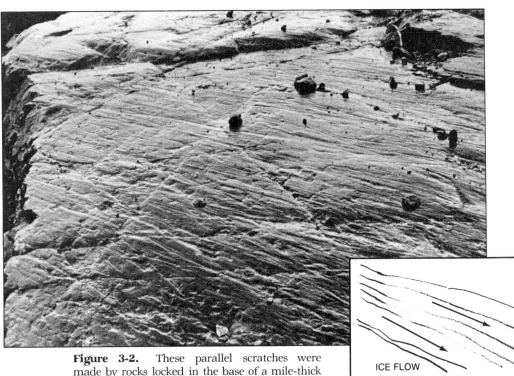

Figure 3-2. These parallel scratches were made by rocks locked in the base of a mile-thick ice sheet that covered this area near Leenane, County Galway, Ireland.

ICE FLOW

Figure 3-3. We are looking down a glacial valley in Nepal. Linear banks of jumbled rocks, existing far down the valley below the ice limit, tell a story of a colder climate in the past.

Figure 3-4. Glacial erratics, such as this distinctive, dark-colored rock resting on light-colored rock, were transported by moving ice. This one is on the Pelham Parkway near the Bronx Zoo in New York City.

colder climatic conditions. The pattern was so clear to him. He saw it wherever he looked. All that was missing was the ice that once occupied the lower parts of the Alpine valleys. Everything else was there.

The old explanation was thought to be sacred because it might be linked to Noah's Biblical flood or, at the very least, to some more natural, high, still stand of the sea. Icebergs with many types of rocks frozen in them were imagined to have drifted across the continent, dragging bottom and scratching the sea floor. Proponents of this explanation pointed to the vast deposits of jumbled rocks, called diluvium, which are spread across the plains of Europe as proof enough that the ice-rafting mechanism had occurred. In this view,

Figure 3-5. Lake basins such as these above the main lake at Killarney, County Kerry, Ireland, were dug out by moving glacial ice.

such deposits were simply the result of the icebergs melting away far inland and dropping their rock loads one by one. The idea of deposition from melting icebergs isn't totally fanciful. Rocks are common constituents of icebergs, and they are dropped onto the sea floor when melting occurs. De Charpentier's audience at the Lucerne meeting knew this and reacted to his paper with scorn and mockery. Even Louis Agassiz opposed the new idea. It was simply too astonishing to be accepted.

Jean de Charpentier was surprised at the resistance offered to an idea that even the country people accepted as perfectly obvious. On the way to the meeting he had met a woodcutter, and they walked along together for some distance. Eventually, the woodcutter remarked, "There are many stones of that kind around here, but they come from far away, from the Grimsel [mountain and glacier] because they consist of granite and the mountains around here are not made of it." De Charpentier was pleased with such a profound statement and asked him how he thought these granite boulders had reached their present locations. The woodcutter answered without hesitation, "The Grimsel glacier transported and deposited them on both sides of the valley because that glacier extended in the past as far as the town of Bern, indeed water could not have deposited them at such an elevation above the valley bottom without filling the lakes."

Jean de Charpentier was so delighted that he gave the old man some money to drink a toast to the memory of the Grimsel glacier and to the preservation of the boulders. Fate takes odd turns. At the time of this conversation de Charpentier was carrying in his pocket a manuscript that would formalize all that the old man had said and bring on a revolution in geologic thought.

Two years later in the summer of 1836, Louis Agassiz decided to spend his vacation at Bex at the foot of the spectacularly beautiful High Calcareous Alps. Jean de Charpentier was working there as director of the salt mines; so an expedition was planned to study the glacial record. Agassiz, de Charpentier, and Ignace Venetz, who was a geologically oriented highway engineer and glacial pioneer, spent several weeks together looking at critical outcrops. The result was that Agassiz abandoned his support of the iceberg hypothesis and became a missionary of the Ice Age. Albert V. Carozzi, a historian of science, describes Agassiz's conversion in these words: "With his power of quick perception, his unmatched memory, his incredible ability to classify and interpret facts, Agassiz assimilated in a few weeks the whole mass of irresistible arguments patiently collected during seven years by Venetz and de Charpentier." Agassiz was on his own and ready to advance.

In November, when the new school year began, Agassiz returned to Neuchatel and began to look for glacial features in the nearby Jura Mountains. The story was the same. Ice action, unknown in the area today, had left its mark of a fairly recent climate change. Karl Schimpler, a close friend of Agassiz, coined the name *Eiszeit* ("Ice Age") and the hypothesis became a theory, expanded from the north pole to the shores of the Mediterranean in a wild burst of the young man's imagination.

Eighteen hundred and forty was a banner year. Agassiz published *Etudes sur Les Glaciers* (*"Studies on Glaciers"),* a major book of eighteen chapters and magnificent engravings. The dedication on the flyleaf is important, because in it we see a warm human being as well as an aggressive opportunist taking advantage of a splendid situation:

> To I. Venetz, Highway and bridge engineer
> of the canton of Vaud
> and
> to J. G. de Charpentier,
> director of the Mines of Bex.

> Gentlemen,
>
> Your interesting works have inspired me to study the glaciers of the Alps. I owe you the first leading ideas which have allowed me to continue these investigations. Therefore, as soon as my observations seemed to me worthy of publication, I decided to dedicate this work to both of you. Please accept it today as a testimony of my deep consideration and affection.
>
> L. Agassiz

Chapter eighteen, entitled "Proofs of the Existence of Large Ice Sheets Outside the Realm of the Alps," was the grand one. In it Agassiz emphasized that scratched-bedrock exposures, lake basins, erratic boulders, and a range of deposits indicated blanketing ice sheets on the continental scale had covered northern Europe and the

Figure 3-6. The edge of the Greenland ice sheet is thin where it is melting back, exposing the jumble of rocks once locked firmly in the lowest ice. The continental ice sheets that covered northern Europe and North America looked like this.

British Isles. The model for this action was the Greenland ice sheet with what we now know to have a maximum thickness of about two miles. Ice in the central section is too heavy to be able to hold its shape and therefore spreads out toward the edges (see figure 3-6) in about the same manner that a block of soap might spread out under the weight of an elephant. Imagine the intellectual daring of a person who could look at parallel scratches on the bedrock in central

Bavaria and then point out the positions of a vanished ice front in one direction and a 1,000-mile-wide, 2-mile-high mountain of ice in the other. Revolutionaries take risks!

"Whatever the opposition against the ideas discussed in this work," Agassiz wrote, summing up his adventure, "it is unquestionable that the numerous and new facts I have presented, particularly with respect to the internal conditions of glaciers, to their action on the substratum, and to the transportation of erratic boulders, have completely changed the *context* in which the question has been discussed up to the present." (italics added)

Agassiz came to America in 1846 to test his theory of an ice age in another part of the northern hemisphere. A beginning had already been made there by two Americans who had found evidence of glacial action in western New York State and Massachusetts. By 1849 Agassiz had become a professor at Harvard University and had discovered the significance of the Great Lakes as glacially carved basins of immense size. Nothing seemed to be beyond the range of his imagination.

Henry Wadsworth Longfellow sensed this and expressed it in a romantic poem written on the occasion of Agassiz's fiftieth birthday party.

> And he wandered away and away
> > With Nature, the dear old nurse . . .
> So she keeps him still a child,
> > And will not let him go. . . .

Charles Darwin

Charles Darwin (1802–1882), author of *On the Origin of Species,* was a completely different kind of revolutionary. Today we tend to think of him as a withdrawn graybeard, fifty years old, or more, who suddenly astounded the world with an iconoclastic view of mankind crouching indifferently among the beasts. That view is partially right. With the publication of his book in 1859, Darwin did alter the course of human understanding; Adam and Eve were supplanted as the first humans, the Bible, overriden as a scientific text. None of these points troubled Darwin a great deal. He had written simply a study of the facts of organic evolution as he saw them and of a process he called natural selection that helps bring the evolutionary changes about.

An anomaly triggered the irrepressible crisis of contradiction that convinced Darwin that organic evolution must be a reality. The story begins with a hoax that led the twenty-two-year-old scientist to appreciate the meaning of anomalies.

Darwin was an indifferent student of both medicine and theology at the University of Edinburgh and at Cambridge before drifting into the natural sciences. In the summer of 1831 he was fortunate to be working as a field assistant to Adam Sedgwick, one of the great geologists of all time. They were mapping rocks in Shropshire, just east of the Welsh border in central England, when a local wag gave Darwin a large, tropical, marine snail shell (*Voluta*), claiming that it came from a gravel pit near Shrewsbury. Darwin thought that Sedgwick would be delighted. On the contrary, Sedgwick declared the shell a hoax: the idea of a tropical sea covering England in recent times was contradicted by too much other evidence. This was the first time that Darwin realized that "Science consists of grouping facts so that general laws or conclusions may be drawn from them."

In August 1831, still working with Sedgwick, Darwin received an invitation to sail on H.M.S. *Beagle* as naturalist without pay for a five-year, forty thousand-mile voyage of discovery. His job would be to observe and to collect everything he could for the British Museum and the expanding concerns of science. He sailed from Plymouth on the twenty-seventh of December with a copy of Charles Lyell's revolutionary textbook, *The Principles of Geology,* in his traveling library. (The second volume of *The Principles* reached him by mail in Montevideo in 1832.) These two books taught Darwin how to read the literature of the earth in the language in which it's written: minerals, fossils, rocks, structures, and land forms. The books also gave him other important perspectives. One was of the vastness of geologic time. A second was the ability to make least-astonishing judgments about the origins of things using Lyell's principle of uniformitarianism. A third may have been the most important of all: The ninth chapter of volume one of *The Principles* is devoted to a negative discussion of the "theory of progressive development of organic life." This may have been the seed that grew into *On the Origin of Species:* if so, the growing was nurtured by the gift of vast geologic times which Lyell gave Darwin.

Darwin's destiny at the moment of sailing on the *Beagle* was predicated on a number of factors: he had the tools with which to work, he knew he had the tools, and the whole world was before him as his laboratory—and he knew that, too!

The climax of the voyage came in September and October of 1835 when Darwin landed on the Galapagos Islands about 650 miles west of Ecuador. An irrepressible crisis of contradiction appeared when he discovered that the ground finches living on the different islands in the archipelago could be distinguished by the shapes of their bills. His journal contained this note: "I have stated that in the 13 species of ground finches a nearly perfect gradation may be traced, from a beak extraordinarily thick to one so fine it may be compared to that of a warbler."

This set of gradational variations shocked Darwin as a potentially devastating anomaly that could refute the established idea of species. At that time the traditional view of species included the assumption that all individuals were alike in every respect. Yet here he was on a small group of islands, hundreds of miles from the nearest continent, seeing variations among individuals rather than conformity to a standard. Something was wrong.

Now, watch the creative process unfold as Charles Darwin thinks his way from finches to Adam and Eve.

The first step was to accept the fact that there was nothing wrong with nature. The error lay in the presupposition that species were created spontaneously in a fixed form and then lived out their

Figure 3-7.   Darwin's *Journal of Researches of the Voyage of the Beagle* shows the range of shapes of the bills found on various ground finches in the Galapagos Islands. Imagine such simple data leading to such profound consequences.

lives without change. The Galapagos ground finches proved that idea to be incorrect. Obviously, species are involved in some sort of continuously unfolding creation. Darwin saw it happening, but he didn't know why it was happening. That point troubled him for years.

A decade later, in 1845, an additional note appeared in the second edition of the *Journal of Researches of the Voyage of the Beagle:* "Seeing this gradation and diversity of structure in one small, intimately related group of birds, one might really fancy that from an original paucity of birds in this archipelago, one species has been taken and modified for different ends. . . . " This note tells us that

Darwin was on the track of evolution and was searching for a mechanism to create modification of species within a small, isolated breeding flock. Several more years passed before he found an answer in the process of natural selection.

Natural selection is a profoundly simple idea. Every animal and plant lives in an environment that both supports it and exerts pressure on it. Rain, drought, heat, cold, predators, disease, variations in food supply, and accidents are typical of the sorts of pressures that may shorten the life span of any individual. Some may be strong enough to meet these pressures and still survive, but those that are less fit may lose out and die off before they are able to pass their weak characteristics on to the next generation. The result of this is a drift in the characteristic form of each generation, particularly if the breeding population is small. That drifting change is organic evolution. Each of us is part of it, just as our ancestors were. The whole story is enormously complicated in detail, yet remarkably simple in general outline.

In Darwin's time practically nothing was known about heredity and the science of genetics. Natural selection may have seemed a mystical idea in 1859, but not today. We now recognize organic evolution to be primarily a genetic story rather than a mechanical one. Much of the resistance to the evolutionary viewpoint has come from people who object on spiritual grounds to the inevitable conclusion that humans are animals, produced by and caught up in the process of change.

It's easy to see why some people were, and still are, upset. A classic anecdote, however, puts their objections in proper perspective. After the burial of Charles Darwin in Westminister Abbey, a worried peer asked Thomas Huxley, one of the pallbearers, "Do you believe that Darwin was right?"

"Of course, he was right," exclaimed Huxley.

His lordship surveyed the abbey with a pained expression, then observed in a low tone, "Couldn't he have just kept it to himself?"

The Darwinian revolution, like the Copernican revolution before it, managed to remove us from the center of the universe. It's probably best to admit that we don't belong there and continue to search elsewhere for our center of stability. One of my friends, a pragmatic auditor, put it very nicely: "I've always made a practice of accepting the universe just as I've found it."

That's exactly what Alfred Wegener (1880–1930) did when he began to accept the realities of something called continental drift.

Alfred Wegener

Wegener was a German physicist with the same level of wide-ranging productivity and unbounded energy that characterized Louis Agassiz. His childhood interests seem prophetically directed toward the consuming goal for which he is now famous. One interest was meteorology, the scientific study of weather and climate. Another was an overpowering fascination with Greenland, the great arctic island that lies buried under a vast sheet of glacial ice. As a child he learned to ski and ice-skate to be prepared to explore Greenland at his first opportunity.

By the time Wegener was thirty-two, he and his brother had held a world record for uninterrupted balloon flight: fifty-two hours. He had earned a doctorate in astronomy, had published original research in the subject, had served as meteorolgist on two scientific expeditions to Greenland, had crossed the mile-high ice sheet on foot, had written several scientific papers on glaciers, basing them on his own field experiences, and had earned a position as professor of meteorology at the University of Marberg near Frankfurt, Germany. Most important, Wegener had written the paper that was to prove the principal springboard for the greatest geological insight since Darwin's.

In this paper Wegener presented substantial proof that our continents have been moved about from point to point across the surface of the earth. In Wegener's model the continents were joined together as a single unit about 300 million years ago and then split up and progressively moved apart throughout the last 180 million years. The idea was labeled "continental drift," with each fragment pictured as a granitic galleon sailing majestically through a sea of basalt. It was a grand vision totally opposed to geologist's long-held dogma of stable continents and ocean basins rigidly fixed in permanent places. How was Alfred Wegener so sure of these startling conclusions?

The answer is that he found an insuppressable crisis of contradiction and decided that continental drift was the only way out of it. His interest in the problem began with an observation that many people had made before him: the outlines of the continents on both sides of the Atlantic Ocean seemed to be complementary. This led him to imagine that he was looking at a gigantic jigsaw puzzle waiting to be solved. That was his great leap of imagination: to see the world as an unsolved jigsaw puzzle rather than as a trite fact. Because of it, Alfred Wegener ranks among the greatest scientists of all time. Even so, his daring idea would have died at birth if it had not been supported by solid evidence that could not be ignored by a glaciologist with field experience in Greenland.

The sting of sleet, wind chill, and frostbite are not forgotten by a geologist who has crossed a major ice field on foot. They are an inseparable part of the environment that includes all the features of erosion, transportation, and deposition produced by glacial ice. Therefore, when Wegener learned that geologists had found massive deposits of glacial rocks, dating from early Permian time (between 270 and 280 million years ago) in South America, Africa, Australia, and India, he immediately reconstructed a glacial climate for those places and times. The shocking part of all this is shown on the map (figure 3-8): Permian glacial deposits cross the equator! Those of southern India lie just north of the equator. Glacial deposits cross the equator in Africa and extend over much of the rest of the land areas of the southern hemisphere with large exposures in South America and Australia. Wegener knew enough about the paleoclimate of Permian time as reflected in the rest of the rock record to be able to rule out the possibility of worldwide glaciation, yet the stinging sleet did blanket areas that are now in the tropics. Here was a crisis

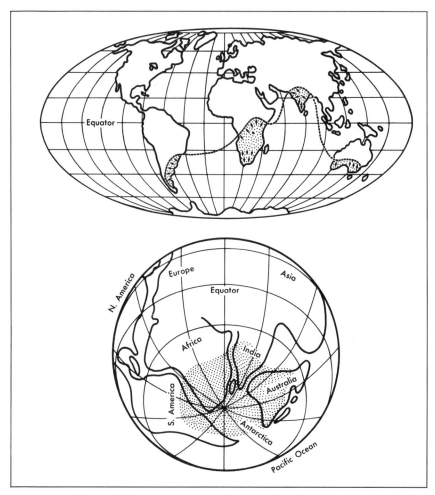

Figure 3-8.   *Top*, Here is the anomalous distribution of the 270 to 280 million-year-old glacial
deposits that triggered Alfred Wegener's scientific
revolution. *Bottom*, The anomaly disappears as
soon as the glaciated land masses are assembled
near the South Pole. (After Arthur Holmes, *Principles of Physical Geology*, 2nd ed., 1965. Courtesy T.D. Nelson and Sons, Ltd., Sudbury-on-
Thames and Ronald Press, N.Y.)

of contradiction that seemed to support only one intellectually daring conclusion: the continents must have been fitted together in such a way that all of these glaciated areas were at the South Pole! There was no other way to account for the facts.

The crisis of contradiction and his solution to it were both strengthened by a knowledge of the directions of ice movement that can be determined by comparing rock types in the glacial deposits with their source areas. That information is shown by the arrows on the map in figure 3-8 (top). There were areas of thick ice that functioned as spreading centers in Africa and India, but ice movements in South America and Australia came from apparently impossible sources, offshore in the oceans. Wegener's solution took care of that by moving the parts of the continental jigsaw puzzle around and eliminating the ocean basins. This is shown in figure 3-8 (bottom). It's a beautiful picture with glacial ice spreading out from centers in Africa and Antarctica over the adjacent parts of Australia and South America. Alfred Wegener's model of continental drift had successfully eliminated the old anomalies but had created a new one.

Where are the forces and machinery that can shift continents from one part of the earth to another? Alfred Wegener never lived to learn the answer. He died in 1930 at the age of fifty while exploring the interior of Greenland, attempting to prove that this northern part of his jigsaw puzzle was originally linked as a land bridge between North America and Europe. Our map (figure 3-8 *top*) shows how well Greenland fits between Labrador and Baffin Island to the west and Norway, Scotland and Ireland on the east. Fortunately, Wegener's geological opinions were published. Other scientists saw the value of his work and were able to carry on. As a result, the 1950s, '60s, and '70s were exciting decades in which the original idea of continental drift was modified. The scientific revolution that began with the young astronomer-meteorologist's recognition of crisis of contradiction became a defensible view of the earth's behavior.

Alfred Wegener was the first person to see and prove that continents move. He was able to do this because glacial rock structures require a glacial climate such as that of Greenland. This insight is in the tradition of Shen Kua, who saw evidence of climate change in fossil bamboo.

I have lived through this revolution and have vivid memories of it. The social forces that operated within the geological fraternity between the 1920s and 1970s reflected the degree of understanding in each part of the community. Very few of us had any true conception of the meaning of a crisis of contradiction. Thomas S. Kuhn's masterful analysis of the structure of scientific revolutions had not been written until the struggle was almost over. As a result, his insights had very little effect on the attitudes that geologists held when faced with the facts about glacial deposits in the tropics. Far too many geologists living in the northern hemisphere simply ignored Wegener's teachings.

We had all been taught the same dogma that continents and ocean basins were permanently fixed in their positions on the face of the earth. It took less effort for those of us living in the northern hemisphere to accept this idea than to defend Wegener against the scorn of our colleagues. After all, no one knew of any force or machinery capable of moving a continent through a sea of rigid basalt. The situation was made more ridiculous by the reactions of geologists living in the southern hemisphere. The majority of them were "Wegenerian drifters" because they couldn't ignore glacial deposits as much as two thousand-feet thick resting on grooved bedrock pavements. These facts were pictured clearly enough in many of the standard textbooks of the day; yet we ignored them.

I cannot remember a single time when I equated glacial deposits in the tropics with driving sleet, a cold climate, and the regional accumulation of a vertical mile or more of ice. To my shame I simply ignored the fact that those glacial rocks and scoured pavements were irrepressible testimony that more snow and sleet fell in the winter than could melt away over the twelve-month-long tropical summers! It's strange to look back on all of this and recognize myself as an impossibly myopic, pseudoscientific clown. That is a confession.

The stories of discovery told in this chapter point out that science is a game of cooperation between the right and left hemispheres of the brain. Input from the dreaming, imaginative, and intuitive right half of the brain is just as important as the logical analysis performed by the left half. The Wegener story shows quite clearly that no success is possible if inept thinking routines block out information. That sort of thing doesn't happen when scientists recognize the high stakes for which they play. Hugh Arthur, a retired aviator, put it very well in one terse sentence, "Nobody cheats in flying school."

Some great poets seem to have an instinctive awareness of the struggle between brain hemispheres. Perhaps it's part of their ability to play for high stakes. Note how the young Percy Bysshe Shelley (1792–1822), who died nearly two decades before the glacial story was finally worked out, sang in "Mont Blanc" of creeping ice and mountains that teach the averting mind:

> Mont Blanc appears—still, snowy and serene—
> Its subject mountains their unearthly forms
> Pile around it, ice and rock; broad vales between
> Of frozen floods, unfathomable deeps,
> . . . .
> Power dwells apart in its tranquillity—
> Remote, serene, and inaccessible:
> And *this*, the naked countenance of the earth,
> On which I gaze, even these primeval mountains
> Teach the averting mind. The glaciers creep
> Like snakes that watch their prey, from their far fountains.

The art of becoming personally aware is simply a matter of controlling the averting mind. It's done by recapturing the importance of a fleeting right-hemisphere insight. Wegener did it when he focused on the potential meaning of the fit of continental outlines on either side of the Atlantic Ocean. There's the real "power that dwells apart in its tranquillity." Capture it, to experience the thrill of discovering context!

Imagine the fun of sitting around a plain oak table and sharing the mellowing glow of a little beer and some great conversation with Louis Agassiz, Charles Darwin, and Alfred Wegener. There's always a tendency for talk to fall to the level of the lowest intellect as a matter of courtesy. It would be a shame to let this opportunity to hear world-class conversation deteriorate because of our own limitations. The thing to do would be to bring up Harrington's First Law of Science— "Nature is scrutable when everything is seen in context"—and get the three of them talking about it. Then sit back and listen to these masters of context explain their views of creativity.

"Hark, Hark! I hear the minstrels play."
Shakespeare, *The Taming of the Shrew*

FOUR

Discovering Context

"Revolution is not a thing you can let others do for you."

MAU KE-YEH

Report from a Chinese Village

To experience the thrill of discovering context all we have to do is focus on something we don't understand and figure it out by using the same thinking processes the great geologists have demonstrated for us. For example, go to a river, stand on the bank, and watch the water as, at a remarkably even pace, it flows by on its way to the sea. A stream exhibits strange behavior when we think about it. The river shown in figure 4-1 is in flood and flowing boisterously, but even so, most of the water particles seem to know their places and stay obediently in line on the long trip to the ocean 240 miles away. That's odd enough to be called an anomaly. What can it teach us about the nature of rivers?

Sometimes the best approach to problems is to break them down into parts small enough to let us see what actually happens from moment to moment. In this case we'll reach into the river and take out a single cubic foot of water. There it is labeled *A* on the left-hand inset drawing. Think of it as a momentarily rigid block of water measuring one linear foot in each direction. Real water won't stand like this for very long without collapsing under the pull of gravity. Therefore, we may expect our water block to collapse at a speed and direction dictated by the acceleration of gravity, the slope of the floor, and the internal strength of the material. It should bulge

79

at the base and spread out in all directions parallel to the slope of
the platform. The water will eventually end up down slope and
nearer the center of the earth.

We need to complicate this picture a little by adding a line of
water blocks such as those labeled *A'*, *B'*, *C'*, and *D'*, in the right-
hand inset drawing. In this situation, block *C'* is confined between *B'*
and *D'* and cannot collapse until there is room for it. That's exactly
what's going on in our picture of the North Pacolet River in Spartanburg County, South Carolina. This is part of a long line of at least 1.5
million water blocks collapsing in lockstep toward the Atlantic
Ocean. All rivers do the same thing. The Ganges, Huang Ho, Nile,
Congo, Amazon, and Mississippi are just long lines of collapsing
water blocks. With this information in context, it's fun to stand on a
riverbank and watch the water blocks give way to one another.

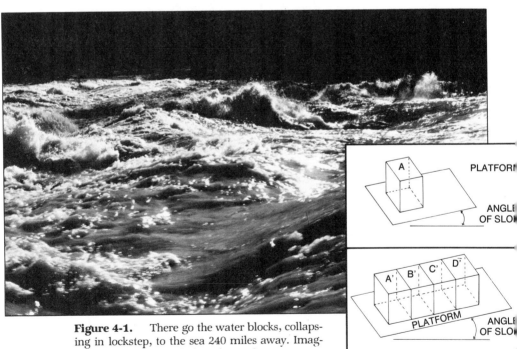

Figure 4-1.   There go the water blocks, collapsing in lockstep, to the sea 240 miles away. Imagine what is going on below the surface.

Look at figure 4-1, and notice the contrast between the smooth, shooting flow of the water in the lower right-hand corner and the wild, splashing, turbulent flow in the background. The different patterns are caused by changes in the character of the channel bottom. Shooting flow, characterized by a laminar, ropy surface, overlines a smooth, steep channel floor. Splashing, turbulent flow reveals the presence of large submerged rocks on which the lines of rapidly moving water blocks impact with great force. Mathematical expressions for this action show that the force of impact against each rock is proportional to the square of the water velocity. Doubling the water velocity increases the impacting force by a factor of four. Unfortunately, the wonder of that marvelous fact is hidden in the frightening language of mathematics. Don't despair; folk wisdom is capable of expressing the same level of wonder in its own idiom.

Our pioneer ancestors were equally descriptive when they named the Kicking Horse River after its exuberant plunge down the west side of the Rocky Mountains in British Columbia. The British poet Samuel Taylor Coleridge (1772–1834) gave us the best picture of all in his description of a meltwater river rushing from the lower end of an Alpine glacier.

> From dark and icy caverns called you forth,
> Down those precipitous, black, jagged rocks,
> Forever shattered, and the same for ever?
> Who gave you your invulnerable life,
> Your strength, your speed, your fury and your joy,
> Unceasing thunder and eternal foam?

Seeing a river in context requires us to look below the surface. Fluid behavior is equally active in the hidden depths where currents move loads of clay, silt, sand, and rocks downhill to the sea. In order to appreciate the reality of these events, we must learn to think of loose sand and pebbles as free particles capable of fluidlike behavior. Consider how easy it is to pour loose, dry sand from one hand to another as if it were a true liquid. A rushing river has power enough to move large boulders in exactly the same way. The results are easy to see in gravel bars and rocky channels exposed to view after flood-waters subside.

Look at the way large stream pebbles are shown to be stacked against one another in figure 4-2. They are set like so many dinner plates tilted upstream. The inset drawing explains how this is done by the impacting force of the moving water blocks. Flat-lying rocks are caught in the current and flipped over into more stable positions where the water pressure tends to hold them in place. Large rocks, weighing as much as twenty to thirty tons, may be stacked in exactly the same way by river currents moving at proportionately higher velocities.

This fact astounded me when I first discovered it about twenty

CURRENT FLOW

Figure 4-2. Look how the river current stacked these rocks on the bed of the River Tilt in Scotland.

Figure 4-3. Rocks with a submerged weight greater than fourteen tons have been neatly stacked by high-velocity flood waters of the Broad River near Chimney Rock, North Carolina.

years ago. The possibility of rivers being able to stack large boulders was something that I had never thought about until the facts were thrust upon me in Hickory Nut Gorge on the Broad River between the towns of Bat Cave and Chimney Rock, North Carolina. Figure 4-3 is offered as proof. Anyone who can recognize the way in which small rocks are stacked in figure 4-2 will have no difficulty seeing the same thing on a much larger scale in figure 4-3. I realized that immediately, but then I was faced with the problem of finding the power to move rocks at least fifteen feet long, eight feet wide, and three feet thick with a submerged weight of fourteen tons.

The answer to that problem was equally obvious. Fourteen tons sounds more formidable than it really is. The rocks are only rolled over so the lifting weight for one edge is merely seven tons more or less. Water moving at a velocity of thirty feet per second could produce enough force against a large surface to tilt the rock quite readily. The problem of available high-velocity floodwater disappears as soon as we look at the steep slope of this gorge.

Here, over a distance of just two and a half miles, the Broad River drops 360 feet down the east flank of the Blue Ridge Mountains. That's equivalent to the total fall of the Mississippi River over a distance of about 850 miles between St. Louis, Missouri, and the Gulf of Mexico. The Broad River doesn't usually look wild. Its bed is so choked with flat angular boulders that the normal river flow is a remarkably quiet passage through pools and down short chutes and runs. However, on rare occasions when Atlantic hurricanes come this far inland and dump many millions of tons of water on the mountains, the river runoff is capable of spectacular power plays. Storm waters running deep and fast have had their way, leaving thousands of rocks tilted together in an impressive record of the whole event. It's a thrilling experience to visit this gorge and realize what you see. Each new level of understanding is an invitation to work out some previously unrecognized problem.

Where did the rocks in Hickory Nut Gorge come from? That's a particularly interesting question because the answer serves as a link with Louis Agassiz and the Ice Age. During the Ice Age, the climate in this part of North Carolina was cold enough to produce a great deal of frost action. Water that filled the cracks in the rocks expanded when it turned to ice. Massive blocks of rock were broken from the one thousand four hundred-foot-high cliff by this process of frost wedging. They tumbled down to partially fill the old, river-cut gorge below. Similar frost wedging remains a common process today, particularly on the north-facing cliff that is never reached by the warming rays of direct sunlight. Accumulations of fallen debris have given the gorge a *U*-shaped profile similar to that of a glacial valley. Broken rocks have filled the original channel to a depth of many tens of feet, and the river is now busily engaged in the task of reexcavating the valley floor. All of these things are waiting to be seen in context by anyone who stops to look for them. Invariably the sight of a partially filled valley brings up the next question: How does a river erode its own valley?

Erosion is beautifully defined by its Latin root words, *e* "out" and *rodere* "to gnaw." How can anything like running water gnaw away hard rock? That problem troubled me for many years until an odd thing happened: I was writing a chapter on the work of rivers for a geology textbook. My fingers were busily tapping the keys of the typewriter when I found that I was thinking of my undergraduate days at Virginia Tech and a game we used to play in the swimming pool with a bowling ball. We were allowed to drop the ball in the deep end of the pool, but were forbidden to drop it in the shallow end for fear of cracking the tiles. I was struck with an amazing insight.

Suppose someone should drop a bowling ball and all of that water into an empty pool at the same moment. Wouldn't that make a pounding splash! That's exactly the action beneath a waterfall. All the rocks, sand, and water that pass over the lip of the falls pound on the river bottom together without any effective cushion to lessen the shock. That was the critical idea. Once I had gotten that far it became perfectly clear to me how to explain how streams carve their canyonlike valleys. Figure 4-4 illustrates the machinery. There is a plunge pool beneath every waterfall where the moving parts—sand, gravel, and water—have hammered out a hole. This is the way streams deepen their valleys. It is also the point where the forces of vertical cutting exceed any tendency for lateral cutting. A short distance downstream from the plunge pool, where the water spreads out and begins to slow down, the bottom is covered by sand and gravel that have come over the falls. All waterfalls and plunge pools

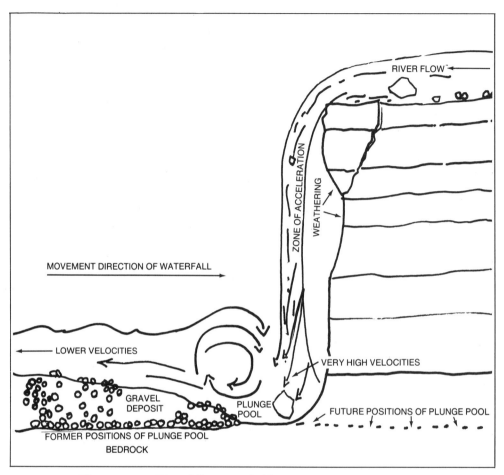

Figure 4-4. Waterfalls are probably the most efficient mechanism for eroding and deepening river valleys. Much of the action occurs on the floor of the plunge pool as rocks of all sizes, uncushioned by the falling water, strike the bottom with high velocities and high-kinetic energies. Crumbling of the lip of the waterfall and weathering of the exposed cliff also contribute to extending and deepening the valley.

move upstream as the headwall weathers back and the resistant lip caves in. This means that waterfalls can be very long-lived. They chase one another upstream from hardrock area to hardrock area functioning as the cutting blades of a great, natural milling machine. In many cases waterfalls probably move hundreds of miles upstream in this manner, deepening the valley beneath them as they move. The 308-foot Lower Falls of the Yellowstone River in Yellowstone Park, Wyoming, is an interesting example. The *V*-profile and the long gorge below the falls tell a story of two movements: vertical cutting and upstream erosion. Notice that the *V*-shaped profile above the falls indicates another waterfall is farther upstream. There is also a wonder story involved.

The Yellowstone River flows northeastward to empty eventually into the Missouri River in western North Dakota. The line of the Missouri River, however, was determined in just the last two million years by the position of the glacial ice front. Meltwater flowed into the Missouri River and off to the Gulf of Mexico. Prior to glacial times, the Yellowstone River probably emptied into the Arctic Ocean by way of the Hudson Bay area. Therefore, the waterfall shown in figure 4-5 may have moved upstream across much of Canada—hardrock area by hardrock area for eighteen hundred miles—long before the time of glaciation and long before any human being was on the continent of North America to see it. How's that for a feat of placing river action in context?

The method is versatile. We can even use it on dinosaurs. It's hard to imagine any geological adventure more thrilling than being able to bring these great beasts out of their lost world and placing them in a living context. I had this opportunity years ago, when I was teaching geology at Southern Methodist University in Dallas, Texas. It was great fun.

My children were young then and we planned a field trip to see the famous dinosaur footprints that were reported to be visible in the bed of the Paluxy River in Somervell County. Apparently the tracks had been made in lower Cretaceous time, some 125 million years ago when the dinosaurs were wandering over the depositional

109 FEET

308 FEET

Figure 4-5.   The 308-foot lower falls of the Yellowstone River in Yellowstone Park, Wyoming, offers a beautiful illustration of a valley deepening by waterfall retreat. Notice the second *V*-shaped valley above this waterfall which was made by the retreat of the 109-foot upper falls about a half mile farther upstream.

surface of the Glen Rose Limestone. Their preservation into modern times was due to river action that constantly exposed new sets of tracks as erosion progressed into successively lower beds of rock. The adventure of seeing these things firsthand offered my children

Figure 4-6. These spots on the bed of the Paluxy River near Glen Rose, Texas, are 125 million-year-old fossil dinosaur footprints.

an opportunity to gain a personal perspective by placing their own bare feet in the same footprints with nothing to protect them except the factor of time.

Our first view of the dinosaur footprints in the riverbed (figure 4-6) was a little scary for the marks appeared to be fresh enough to have been made that morning. We caught ourselves looking around for a place to hide, just in case a hungry sauropod should appear. The children were hardly reassured by my explanation of our four-dimensional world of latitude, longitude, elevation, and time. I pointed out that we all occupied the same spatial coordinates, give or take a little continental drift and uplift, but we were visitors to this place in completely different time frames. Children are practical. Mine soon shrugged away their fears and stripped off their shoes and socks, eager to get into the river.

The white Glen Rose limestone is quite pure and weathers away by solution without leaving a significant residue of sand and clay. As a result, the riverbed is hard and clean with the dinosaur footprints beautifully exposed on the bottom under about eighteen inches of clear water. Although the limestone is now hard, it was a soft, squishy carbonate sand when the dinosaurs stepped on it years ago. Their weight compacted some of it and squeezed the rest out between their toes and up around their heels. Every detail of this can be seen in figure 4-7. Notice the elevated rim around the print. Claw marks are faintly visible at the ends of the toes. Preservation of this much detail is due to selective solution. All of the younger layers of uncompacted carbonate sand that once filled the prints have been dissolved away much more readily than the compressed sand underfoot. This gives the exposed bed the sharp outlines that account for its appearance of having been occupied within the last few hours. We all felt a little like Robinson Crusoe, comparing foot sizes and strides.

My feet are about eleven inches long, and my comfortable stride in water is just under two feet. This compares favorably with the dinosaur's eighteen-inch footprints and five-foot stride. Apparently we shared something of the same general proportions. Evolutionary changes since Cretaceous time have modified the shape of our walking feet, expanding the length of my foot from heel to sole, and shortening the lengths of the toes. One glance will show that our hands are still proportioned much more like those of the dinosaurs. Even though they are not in our direct line of descent, we do share an ancestry with their country cousins. This is obvious to anyone who cares to look at the fossil data. Comparing measurements is a delightful experience, for it brings us closer to appreciating our relationships with all life on this planet.

All of these contexts flashed through my consciousness as I walked happily about in the knee-deep water taking photographs. Suddenly it occurred to me that I was missing something: Why had these dinosaurs been wading about on the sea floor in the first place?

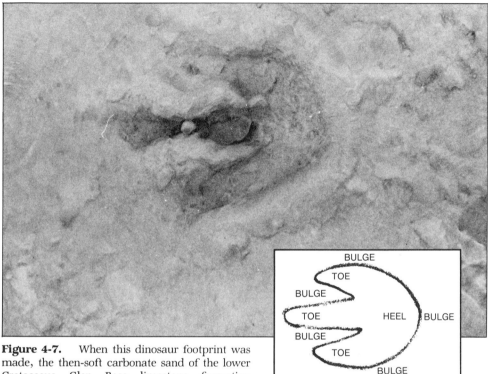

Figure 4-7. When this dinosaur footprint was made, the then-soft carbonate sand of the lower Cretaceous Glen Rose limestone formation squeezed up around the heel and between the toes.

The point was especially interesting to me because I had done some fieldwork on the deposition of similar limestones on the Bahama Banks; I knew what this Texas area must have looked like over 100 million years ago in lower Cretaceous time. Water absorbs visible light at the red and yellow end of the spectrum more readily than it does at the green, blue, and violet end. As a result, watercolors over a white limestone sea floor change with water depth to beautiful pastel greens, blue-greens, and then to deeper blues. At what is now Glen Rose, Texas, the Cretaceous ocean must have been about ten feet deep and tinted a lovely bluish-green during the time of limestone deposition. Perfection of the footprints implied quiet water, perhaps in a protected lagoon with very little wave action. I realized that these ideas were a good start, but I needed more information.

There had to be something else, some clue that would explain why terrestrial dinosaurs were out here wading about in the sea. I looked at the rocks again and was almost staggered by the obvious answer, seaweed fossils! (figure 4-8) My friends had been walking over this carbonate mud flat at low tide to graze on well-salted salads.

I called my family over and we tried to conjure up an image of the herd. Imagine the pleasure that's available to anyone capable of visualizing a great dinosaur rising from the rock, filling her footprints, clothing her fossil bones with flesh and varicolored scaly skin,

Figure 4-8. Look at the seaweed fossils. The good grazing attracted the dinosaurs out on this limestone bank.

stretching her neck, opening her duck bill or spike-toothed mouth and calling to her mate across the estuary. That's living.

*   *   *

Learning to enjoy reality, that which is and was, is a personal revolution that no one else can do for you. Learning at its very best is the discovery and then, at a deeper level, the realization of context. I have pushed the point very strongly in this chapter because I want to make the distinction between what we have been told is true and what we know to be true on the basis of our own experiences and abilities to reason.

PART 2

Reading Rocks: Decoding the History of the Earth

Early Work on the Grand Canyon.

David Macaulay. Courtesy *Atlantic Monthly.*

FIVE

The Wasness of the Is

"How does a play begin?
In medias res!
In the midst of things!"

GEORGE L. KITTERIDGE

*Great Teachers Portrayed by Those
Who Studied Under Them*

Plays begin in the midst of things and so do our views of the earth. A few scientists working quietly since the dawn of civilization have recognized that creation was a matter of progressive change through time. We will call the record of these events "the wasness of the is." Things that were once thought to be just ordinary parts of every existing landscape were discovered to have unique and often well-defined histories. We have seen an example of this in the early work on the glaciers of Switzerland. Jean de Charpentier and Louis Agassiz saw evidence in the valleys which told them of the preceding Ice Age. Remember that the scratched pavements (figure 3-2) exist today. Each exposure is part of the modern world; so we use the present tense to indicate the fact. Seeing scratches was not what made Agassiz famous. He discovered the "wasness" of those scratches: therefore they can also be described in the past tense, indicating the older relationship between rock and time. The idea has beautiful bloodlines. It was transposed

97

from Thomas Mann's "isness of the was," a phrase that referred to ideas carried into the modern world from ancient times. Think of the knowledge of the wasness of the is as a key that will unlock the mysterious history of the earth for us.

All we need to do is learn to use the key. Once we have mastered that, the story of the earth is waiting there to be read in the language in which it is written: in minerals, rocks, fossils, structures, land forms, and processes. Our first step is to understand the time gap between *was* and *is*. We'll begin by learning to recognize the passage of time as a duration between events.

This is a familiar idea. We usually think of a second as the duration between the tick and the tock of a pendulum clock. Physicists demand more precision, however, and define a second as the duration required for 9,192,631,770 ± 20 cyclic vibrations of cesium atoms in an atomic clock. The duration of a day was set from noon to noon because that unit isn't warped by the seasons, as a day defined from dawn to sunset would be. Lunar months and years are also based on real events of standard duration. (All other common time units are convenient artifacts established by counting seconds, days, and years.) One important aspect of our standard time units is that they are all repetitive and exactly alike.

Geologic time units are much more fun to work with because each one is unique. All we have to be able to do in order to discover a span of geologic time is recognize two events, one of which can be proven to be earlier than the other. The duration between them is a time unit. A set of tree rings, such as that shown in figure 5-1, is familiar enough to serve as an introduction to this method of thinking.

I found this block of redwood many years ago. It simply appeared on my front lawn. Since wood carving was one of my hobbies at the time, I picked it up intending to release the lovely wood nymph that obviously was trapped inside. All thought of creativity faded away, however, as soon as I saw the pattern of the grain exposed on the end of the block. This piece of wood was an antique, too venerable to be carved.

The lines in the picture represent annual rings. Each pair, light

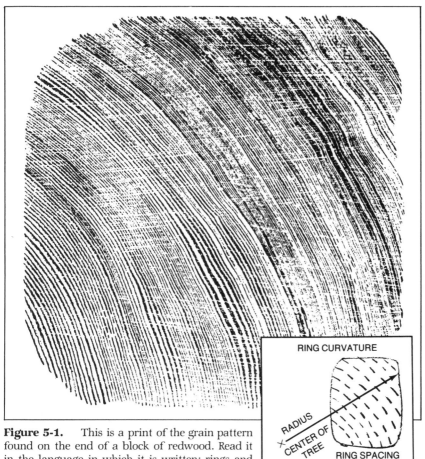

Figure 5-1.   This is a print of the grain pattern found on the end of a block of redwood. Read it in the language in which it is written: rings and spacings.

and dark, represents the amount of growth that occurred in a single year. The direction of curvature reveals the positions of both the center and the outside of the tree. There are 243 pairs of rings. That means that the lower left-hand corner of the block is an antique, 243 years older than the upper right-hand corner. It's possible to use similar reasoning to estimate the minimum age of the whole tree.

We cannot be absolutely accurate because we don't know how

many additional rings may have been cut away by the sawmill. However, just using the record we have of the curvature of the rings, we can estimate the radius of the tree to have been no less than twenty inches. This means that the tree was at least one thousand, perhaps fifteen hundred years old when it was cut down to be made into lawn furniture. Therefore, the block could be a great deal older than that since the radius of a commercial redwood log may be as much as forty-eight inches. An additional twenty-eight inches in radius could represent another eleven hundred years of growth.

Notice how some of the sets of rings are very close together while others are quite far apart. These differences were caused by repeated climatic variations that lasted for the better part of a decade. Botanists have learned that thin annual rings are the result of stress induced by cold winters followed by dry summers. Thick rings reflect growth-producing conditions such as warm, wet winters and moist, yet sunny, summers. This 243-year record of the climatic variations of Northern California suggest that climate change is a normal part of our environment. An even more significant point is that our reasoning was based on the identification and interpretation of things called primary structures.

Primary structures have two distinctive qualities that are important in geology. The first is contained in their definition. Primary structures are features produced in all natural things at the time of origin and as a direct result of the way the parts were put together. The second distinctive quality of primary structures is that they may be viewed as the record of an event that occurred at a certain time. Therefore, when we identify a primary structure, we also identify a moment in that period.

The ellipsoidal shape of a bird's egg is a good example of a primary structure that marks the moment the egg was laid. Forced extrusion through the bird's sphincter muscle imposes the familiar circular cross-section shape on the egg at the moment of separation from the bird. The sphincter's resistance to extrusion results in the asymmetrical elliptical shape of the egg. That streamlined design is much less excruciating than one with points and corners. No wonder it has been used by amphibians, dinosaurs, reptiles, and birds for at least 300 million years.

This is an important point, for a most uncommon amount of common sense led Nicolaus Steno (1638–1687) to the discovery of primary structures and how to use them to interpret the history of the earth. Steno was playing the part of an intellectual Christopher Columbus teaching us to navigate in an uncharted sea of nearly limitless time. He was the right person with the right experience who became aware of the right anomaly and then devoted his considerable energies to the opportunity to make more discoveries.

Nicolaus Steno was a Dane, educated at the universities of Copenhagen, Amsterdam, and Leiden, where he received a medical degree. Most of his early scientific work was in anatomy, classifying things by their shapes, boundaries, and functions. For some long forgotten reason, Steno left a research position in Paris and became attached to the Medici court under the sponsorship of Ferdinand II, grand duke of Tuscany, a province in the northwestern portion of what is now Italy. (Ferdinand had been a student of Galileo's and was an experimentalist in his own right. Early in 1666 fate seems to have taken a hand in Steno's life. He had gone to Ferdinand's winter residence at Pisa at just the moment that some fishermen caught a great white shark and dragged it ashore. The head was given to Steno for anatomical study. His report in 1667, entitled "The Head of a Shark Dissected" contained an insight that became the seed from which his most important scientific work developed.

Steno recognized that highly prized natural curiosities called tonguestones were actually fossil shark's teeth. Once he had convinced himself that the analogy was correct, Steno made a magnificent intellectual leap of reason and imagination that breathed life back into the fossils and pushed him toward an even more exciting search for the historical context of living sharks.

Steno knew that tonguestones are found in the rocks of the Maltese Islands, south of Sicily in the Mediterranean Sea. Sharks are sea creatures, not landlubbers; hence these rocks must have been marine sediments deposited in such a way as to have enclosed sharks' teeth, presumably after death or shedding. Steno was forced to conclude, therefore, that part of the Mediterranean sea floor must have been lifted locally to form the Maltese Islands. (Note the parallels beginning to emerge between the work of Steno in seventeenth-century Italy and that of Shen Kua, who found the fossil bamboos in China during the twelfth century.)

Steno realized that his ideas were profound enough to deserve further testing. Within two years, he had made a number of extended field trips to study rocks and had written a lengthy abstract for a book that he intended to complete later. His title was formidable: *The Prodromus of Nicolaus Steno's Dissertation Concerning a Solid Body Enclosed by the Process of Nature Within a Solid.* This manuscript contains the great generalizations that were to form the intellectual foundations for the science of historical geology. By 1671 Henry Oldenburg had translated *The Prodromus*, as it is known today, from Latin into English and had advertised it to the scientific community in the Philosophical Transactions of the Royal Society.

Steno's ideas were picked up over a century later by James Hutton, who made a career of identifying primary structures, interpreting their meanings in his own treatise, *The Theory of the Earth.* Forty years after that, Charles Lyell developed the *Principles of Geology* on exactly the same foundation: primary structures, reason, and imagination. And now, two hundred years after Lyell, we will have similar adventures by learning to interpret primary structures ourselves. Our purpose is to read the literature of rocks in the language in which it is written.

Our technique will be to visit a series of outcrops, examine them in detail, and then interpret the stories they have to tell us. I'll begin by presenting a little background to explain where we are and put the place in context. After that we'll begin to look for primary structures with the systematic eye of an anatomist. We'll observe details of form and structure. Each significant unit will be identified by the boundaries that separate it from other things. Once we know what we're looking at, we'll play Sherlock Holmes and make least-astonishment value judgments that will serve to explain how the parts are related to one another in space, function, and time. The fun will come when we begin to distinguish the separate ages of things, that is, when we find ourselves able to wrap each piece of geologic history in its own particular "time-bag."

It is possible to think of time-bags as entities with visible boundaries. Each annual ring in our redwood block, for example, belongs in a separate time-bag, and we can see the divisions between them precisely. Each is a primary structure formed at a unique moment. Most of the mysteries of geologic history begin to fade as soon as we learn to look for time-bags and the boundaries between them.

Figure 5-2 pictures a wonderful exposure of a glacial deposit. My wife and I stumbled upon it while exploring the north side of the Dingle Peninsula in County Kerry, Ireland. We had left the famous town of Tralee, driving west between the Sleive Mish Mountains and Tralee Bay. A few miles beyond Rough Point, a narrow track led us down to the shore of Brandon Bay between Stradbally and Kilcummin. (Ah, these wonderful Irish names!) The beach sand there is the same spectacular reddish-brown as the source material, the old red sandstone of Devonian age (i.e., 360 million years old) which makes up the backbone of the peninsula. The tide was low, and we ran hand in hand across the wide, shimmering beach, making two sets of unusually deep and sharply cut footprints. After wetting our feet in the sea, we turned back toward the mountain and a fifteen-foot-high sea cliff to find a dry spot for a picnic. Suddenly my whole frame of reference changed, and I became lost in time. Storm waves had exposed a rock to cheer the heart of Nicolaus Steno (figure 5-2).

The Prodromus contains an observation that any solid that possesses a boundary already must have become hard when the matter around it was still capable of behaving as a fluid. We can look at the cliff and identify boundaries that surround every boulder, every pebble, and every grain of sand. Each is distinct because it has an outer surface, just as we are distinct within our skins. Steno saw these "skins" as time-bags with dates on them. In this case, the dates refer to the time of glacial erosion that broke the materials from the

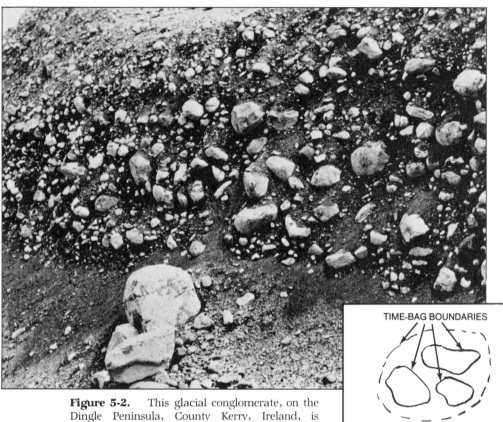

TIME-BAG BOUNDARIES

Figure 5-2.   This glacial conglomerate, on the Dingle Peninsula, County Kerry, Ireland, is younger than its parts.

mountains to the south and east. Of course, the rocks inside these time-bags are a great deal older than their dates of glacial erosion.

Sometime between fifteen and twenty thousand years ago, all of the local glacial ice melted away and left this jumbled conglomerate as a record of the event. Imagine one plastic time-bag several miles long and several miles wide enclosing this entire deposit. The bottom of it lies on bedrock. My wife and I had driven across the top to get to the beach. One side is the surface of the sea cliff. The other sides are of irregular shape and connect the bottom to the top. This time-bag represents a primary structure formed by a single tick and tock with a duration of some part of the Ice Age in between.

This method of dating the different parts of a rock is Steno's first great contribution to the science of geology. (The idea works just as well on the mineral crystals of igneous and metamorphic rocks as it does on the pebbles, grains, and fossils found in sediments.) One of the best ways to demonstrate Steno's truly remarkable insight is to apply it to rocks that contain materials of several different generations.

The idea of multiple generations of rocks is simple to grasp. Each new rock is compounded of the weathered debris of older rocks. This series of events is commonly known as *the rock cycle.* A good, though grisly, analogy is that of a community of cannibals who live through the centuries by eating one another. The example of multiple rock generations that I found in the sea cliff on the Dingle Peninsula represents an epic of heroic proportions. I'm reminded of the first lines sung by the bard of Beowulf: "I shall unlock my hoard of words at your will. I shall unfold before you, O King, the story of my wanderings, for truly, I have wandered long years over this wonderful world. . . . Only the great in heart [are] worthy of being sung by the bards. . . ."

I found the four-inch pebble shown in figure 5-3 because I was looking for it. Hunters are attuned to specific patterns. A turkey hunter doesn't look for birds; he looks for scratches among the leaves on the forest floor that tell him turkeys are near. Pattern recognition is an ancient, gene-imprinted trait. On this occasion I was set to look for boundary lines of smaller rocks enclosed within a larger one. The best specimen I found contains eight separate primary structures that tell of four successive generations of source rocks. Count them as we wrap each one in its proper time-bag. We'll work our way back in time beginning with the youngest record of an event and ending with the oldest. This approach forces us to begin at the outside of the rock and work our way inward. Remember that during the time it was being formed, each part of the rock was capable of behaving like a fluid, just as loose sand poured from one hand to another behaves as a fluid. Don't be fooled by the fact that the rock is hard now. That's a later story involving welding and cementation of the grains.

Under these rules, time-bag *M* encloses the whole pebble and is known to date from the Ice Age and the time of glacial erosion. Its

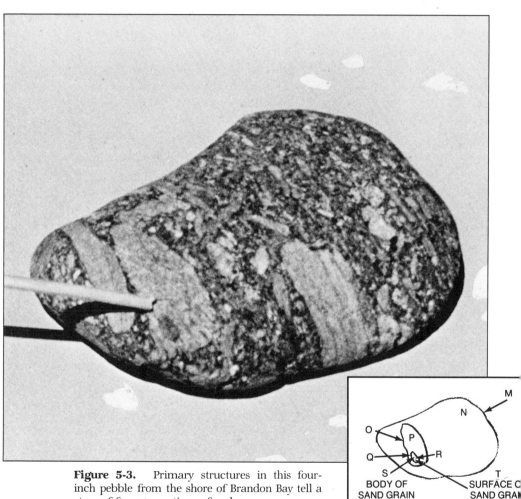

Figure 5-3.   Primary structures in this four-inch pebble from the shore of Brandon Bay tell a story of four generations of rocks.

age could be anywhere in the last two million years, but not younger than about twenty thousand years. The reddish-gray rock within this time-bag was easily identified in the field as a recognizable piece of the Old Red Sandstone. This type of rock outcrops on the Dingle Peninsula and in many other parts of Ireland, Scotland, and England. The Old Red is a sediment deposited from middle to upper Devonian time, making 360 million years a fair estimate of its age. We'll put the Old Red in time-bag *N*, accepting an *N*-to-*M* interval as about 360 million years.

Look just below the end of the pointer at the elongated pebble with the rounded corners. It has a boundary surface that we'll place in time-bag *O* to mark the moment it was freed by erosion from some large mass of source rock. A piece of this light colored material is displayed within *O*; we'll call its time-bag *P*. More exact dates for *O* and *P* are unavailable. However, event *P* does contain more information.

Notice the small, dark colored, oval pebble within *P*. It is the dark spot just below the end of the pointer. Its surface is surrounded by time-bag *Q* marking the moment of erosion when this pebble was separated from some more ancient source rock, with an age represented by time-bag *R*. This piece of *R* became free in time *Q* and was collected under conditions of fluid behavior to be enclosed with other materials in the pebble of time-bag-age *P*. We know nothing of the absolute ages of these early events, but are certain of their relative positions in time and can continue to search for a beginning.

Another piece of evidence has been waiting a long time for us to find it. Look at the light spot marking the position of a grain of quartz sand in the lower right-hand corner of pebble *R*. This grain was also freed by an erosion event we'll put in time-bag *S* to distinguish it from the date of the quartz source rock that belongs in time-bag *T*, the oldest event recorded in figure 5-3. Who knows how much real time is marked out with these sets of letters and time-bags! The story is well fitted to Shakespeare's lines from *The Tempest:*

Tis far off.
And rather like a dream than an assurance
That my remembrance warrants. . . .

This approach is a good deal like looking into the parallel wall mirrors of a barbershop and seeing the images fade toward infinity. Our four-inch pebble contains the primary structures of four generations of erosional events and four generations of rocks. Listed from oldest to youngest, the rock events were placed in time-bags *T, R, P,* and *N,* and the erosion events in time-bags *S, Q, O,* and *M.* A similar approach could be used on the large glacial deposit for it can be considered as a large, flat superpebble for the purposes of time-bag analysis.

This example is enough to show why Nicolaus Steno is considered the father of historical geology. Yet, fortunately, there is more. Steno also taught us something now called the law of superposition which states that every set of undisturbed sediments is arranged in chronological order with the oldest bed on the bottom and the youngest on the top. This contribution started geologists on the path toward developing a reliable geologic time-scale. It also led to the science of making geologic maps for entire continents and similar maps of the moon and Mars.

It's not surprising that Steno, the anatomist, would focus his attention on a major structure as obvious as the parallel layering of sedimentary rocks. These layers, or beds, have upper and lower boundary surfaces; each, therefore, belongs in its own time-bag. Once Steno recognized this fact, it was an easy, and logical, step for him to point out that the oldest is at the bottom and the youngest at the top. Think about it for a moment: loose sand and silt are capable of behaving as a fluid. For such material to be spread out by water or air currents in a thin layer over a large area, it must have had something older and firmer to rest upon, that is, a bottom layer solid enough to support an overlying layer. The same reasoning applies to succeeding layers of decreasing age from bottom to top.

SEDIMENTARY STRUCTURES MAKE SENSE

With this as a background, examine figure 5-4, showing a thin-bedded, shaley rock outcrop on the flank of the Smoky Mountains in eastern Tennessee. There are many thousands of sets of paper-thin beds and time-bags in this exposure. Imagine that is Steno's finger pointing to a single unit. Hear his voice emphasizing that this material was deposited as a thin, soupy, clay-mud falling gently down through quiet seawater to the deep ocean floor far from shore. Thousands of nearly identical layers in this place imply that similar floods brought clays from the continent. Events such as this do not occur every day or every year. Rapid deposition, without consolidation be-

Figure 5-4.   This is a multilayered, shaley deposit near Townsend, Tennessee. The pointing finger covers five or six thin layers.

.tween layers, would not furnish the solid floor that is clearly defined at each bedding surface. There may have been decades of compaction after each episode.

Imagine the awe with which my students viewed this exposure after counting the number of layers per inch and looking at the great height of the road cut before us. Here is an example and an experience as impressive in its own way as that gained by applying the law of superposition to the wall of the Grand Canyon. Once we have reached this level of appreciation, we are ready to thrill at Steno's next great insight: the principle of original horizontality of sediments.

The principle states that sedimentary beds are usually deposited with their bottoms conforming to the previous shape of the basin and their tops nearly parallel to sea level. This is almost self-evident. Sediments prior to compaction behave to some degree as fluids. This tends to level the fine-grained clay mucks under the pull of gravity. There is also a marked tendency for the sediments to be deposited in low places so that they are filled to the same level as the rest of the unit.

The importance of Steno's principle of original horizontality is that it gives us a reference level against which we can judge what postdepositional changes have occurred. Consider, for example, the great plains of the American Midwest, classically shown in parts of Texas, New Mexico, Oklahoma, Kansas, Nebraska, and eastern Colorado. Much of the flatness of these landscapes is due to the original low slopes and near horizontality of the sedimentary bedrocks. These things tell us that there has been very little distortion of the continent since the time of sedimentation. The only major movement of the continent recorded here is a broad regional uplift that brought marine sediments out of the sea. The same situation is found on most of the coastal plains of the world. A beautiful example is seen in eastern North America from Long Island, New York, to Mexico's Yucatan Peninsula. In places, this strip of flat-lying beds is hundreds of miles wide, accounting for the topography of Florida, Mississippi, Louisiana, and South and East Texas. Contrast these situations with the structure of the Sangre de Cristo Mountains in southcentral Colorado (figure 5-5).

Rocks have complex histories of both formation and deformation that can be distinguished by their field relationships. Steno never said this in so many words, but he showed that he understood the principle by drawing cross-sections of what he believed to be the structural history of Tuscany. He contrasted horizontal and tilted rocks and recognized that tilting movements followed the deposition

Figure 5-5. Sedimentary rocks have been folded into a long inverted *U*-shape in this part of the Sangre de Cristo Mountains of Colorado. There is no way for sediments to be deposited in this form.

of horizontal strata. We can say the same thing about this great fold in the Sangre de Cristos. Notice how the beds stand on end on the right (east) side of the range, bend sharply over the crest, and turn down again on the left (west) side. It's impossible for a great thickness of sediments to be deposited in this manner. The only interpretation that makes sense is that folding is an event that belongs in its own time-bag. According to the law of superposition each layer of sediment must be contained in an individual time-bag. Before we can bag these beds, we must flatten them out again as they were originally deposited and see them in their chronological order.

Steno certainly had a good imagination. He was the first to recognize the meaning of crosscutting relationships such as the one shown in figure 5-6. In this type of deformation, a once-liquid rock has been caught in the act of invading a limestone. This example is exposed beside a busy street on the mountain in the heart of Montreal. A geologist looking at this Canadian outcrop sees a 420 million-year-old, mid-Ordovician limestone invaded by a 100 million-year-old, Cretaceous granitic rock. James Hutton's debt to Steno is reflected in this structure for it is analogous to the outcrop at Glen Tilt, Scotland. All we need to do to complete the picture is to add the time-bags.

Notice how the stack of horizontal, 420 million-year-old time-bags that enclose each of the limestone beds has been broken and separated. This gap is filled in with a set of three vertical time-bags. One of them is 100 million years old and encloses the granitic intrusion. The other two are wedged into the cracks separating the left and right sides of the intrusion from the limestone. Each of these time-bags contains 320 million years. This is the interval between the formation of the limestone and the time of intrusion. Accountants should enjoy this phase of geology for the time sheets always balance.

It's difficult to imagine how much more Steno might have been able to discover if he hadn't become a priest, preoccupied with other duties, including those as Titular Bishop of Titiopolis, a part of modern Turkey. Nevertheless, he gave us a rational approach to the study of rocks and left the rest to those who came after him. As apprentices we're all included in that group.

Figure 5-6. This outcrop in Montreal displays 420 million-year-old, mid-Ordovician limestones that have been cut by a 100 million-year-old Cretaceous igneous rock.

The time has come to test our skills on new problems by looking at a few photographs of field situations and seeing how well we can read the geology for ourselves. We'll begin at Monument Valley on the common border between the eastern parts of Arizona and Utah. Deserts are wonderful places to see geology because there is so little cosmetic cover of plants and soil to hide the rocks.

The key to understanding Monument Valley lies in correlating the sedimentary layers from one pillar to another. A set of lines has been drawn on the picture to demonstrate what the word *correlation* means. Notice that there are three identical units in every monument. Each belongs in its own time-bag. Start at the base with the flaring support formed by 250 million-year-old, mid-Permian sandstones and shales. Above that there is a vertical, cliff-forming unit. This is made of a beautiful sequence of well-cemented, 245 million-year-old, mid-Permian desert sand dunes. The pancake-like layers on top are composed of mid-Triassic sandstones and conglomerates ranging in age from about 215 million years at the bottom to about

Figure 5-7.   Monument Valley is a great place to look for horizontal time-bags that have been cut up by modern erosion. These layers of rock extended laterally for many miles in every direction until about 10 million years ago. Can you see rocks that are no longer there?

200 million years at the top. Obviously, the monuments are remnants of a sheet of rock that once extended for some distance in all directions.

Dear old Steno saw the meaning of this sort of thing long ago and gave us a generalization to cover it. His insight is sometimes called the principle of continuity: when we see the edges of beds, we may be sure that something has been removed to expose them. That's certainly true at Monument Valley. We can see the raw edges of the layers and know that a great deal of rock has been removed. We can even tell how much rock has been removed because the distances and heights of the pillars are well-defined. Regional geologic maps show that the erosion extends for many miles beyond the last pillar. Suddenly these monuments, which seemed indestructible at first, have become mortal, the last remnants of a golden age of greater grandeur. We're reminded of Percy Bysshe Shelley's "Ozymandias."

> I met a traveler from an antique land
> Who said: Two vast and trunkless legs of stone
> Stand in the desert. Near them, on the sand,
> Half sunk, a shattered visage lies, whose frown,
> And wrinkled lip, and sneer of cold command,
> Tell that its sculptor well those passions read
> Which yet survive (stamped on these lifeless things),
> The hand that mocked them and the heart that fed.
> And on the pedestal these words appear:
> "My name is Ozymandias, king of kings;
> Look on my works, ye Mighty, and despair!"
> Nothing beside remains. Round the decay
> Of that colossal wreck, boundless and bare
> The lone and level sands stretch far away.

Nothing is permanent; yet there is a brighter side: geology is not just the story of destruction. There are building processes at work, too. Erosion takes rock away, but in the ever-changing and continuous pattern of creation it sculpts beautiful things as well.

Figure 5-8.   Giant waves such as this one are
not uncommon on the shores of the Aran Islands
west of Ireland. This cliff is about sixty-five feet
high and the wave has splashed at least twenty
feet higher than that. Think of the striking force
that must be absorbed every few seconds by these
rocks.

The Aran Islands off the west coast of Ireland are exposed to
strong, buffeting winds that sweep inland from the open sea and
force everything to adjust to their power. There are no trees. Living
things hug the ground and seek protection behind a net of crisscross-
ing walls. Thatched roofs must be tied down, and salt spray is a
familiar taste in the air. Wave action on the boundary cliffs is often
astounding. The example shown in figure 5-8 cannot have splashed
less than ninety feet into the air; yet the day was bright and the sea
seemed reasonably tame. Imagine the force of tons of water moving

at vertically deflected speeds as high as seventy-five feet per second. It's easy to see how such great waves could lift rocks from the ocean floor and hurl them like artillery shells against the cliffs. Naturally, the land becomes dented, parts of it left behind as promontories, isolated rocks, and islands.

Intermediate stages of shoreline development are familiar sights along these rugged coasts. A typical, small-scale example from the northwestern part of Scotland is shown in figure 5-9. Little imagination is required to reconstruct the story by filling in the eroded areas with their original materials. A good deal more imagination, however, is necessary to see larger-scale features for what they are.

The Aran Islands are monuments much like those at Monument Valley. Broken edges of the 330 million-year-old lower Carboniferous limestones of which they are built indicate correlation with similar rocks many miles away on the west coast of the mainland. Addi-

Figure 5-9.   This stretch of rugged coastline is in northwestern Scotland. Compare what you see to the land forms of Monument Valley, Utah. The isolated rocks and islands in this picture are also monuments, temporarily surviving the forces of erosion. Restore for yourself the original continuity.

tional correlations across the width of the Irish Sea link the rocks of
Ireland with those of Scotland. Therefore, Ireland itself is a monument, too. Correlation may be made from Great Britain to the mainland of Europe, indicating that this island country is also a monument separated from its roots by a combination of river and marine
erosion. The important point is to learn to use the word *monument*
in two ways simultaneously. Islands are monumental land forms
standing above the sea. They are also monuments left in memory of
the great events that carved them from their older source rocks. We
know that Shakespeare had some grasp of this because the 64th
Sonnet contains these lines.

> When I have seen by Time's fell hand defac'd
> The rich-proud cost of outworn buried age;
> When sometime lofty towers I see down-ras'd,
> And brass eternal, slave to mortal rage;
> When I have seen the hungry ocean gain
> Advantage on the kingdom of the shore,
> And the firm soil win of the wat'ry main,
> Increasing store with loss, and loss with store;
> When I have seen such interchange of state,
> Or state itself confounded to decay;
> Ruin hath taught me thus to ruminate -
> That Time will come and take my love away.
> This thought is as a death, which cannot choose
> But weep to have that which it fears to lose.

Only a field observer could have written, "When I have
seen. . . ." Grant that Shakespeare was an observer, what could he
have meant by the lines: "And the firm soil win of the wat'ry main/
Increasing store with loss and loss with store"? The answer is found
in figure 5-10.

This picture was taken on the north coast of Ireland in County
Donegal, where great stretches of sandy beach are made from materials broken from the nearby cliffs. Land is built out into the sea at
these points. This is a very stable pattern because, as you can see, the

Figure 5-10.   Wide, sandy beaches such as this one in County Donegal, Ireland, wear out the waves as they move uphill. Here we see, in Shakespeare's words from *King Henry IV*, "The beachy girdle of the ocean/Too wide for Neptune's hips."

wave energy is absorbed harmlessly. Part of this is due to frictional drag on the sea floor and part by the effort of lifting water onto the sloping sand. Notice how the waves become worn out as they move onto the beach. New sand is added as older materials are worn down. It's a beautiful story and part of the same theme we uncovered at Brandon Bay on the Dingle Peninsula while discussing the rock cycle. Almost every part of the wasness of the is weaves together if we look for the connections.

In figure 5-9 there is a flat meadowland about fifty feet above the sea on our side of the far rocky point. Features such as this are found in many areas once buried by the continental ice sheet. They represent marine benches beveled by wave action and lifted above the sea by a type of rebound that follows the removal of the ice load. Notice that part of the old sea cliff is still visible. Similar beveling action is taking place below sea level today. The earth is slow to rebound, and we may expect another terrace or two to be cut and lifted before the final effects of glaciation are completed. In some places along these coasts it is possible to identify as many as three terraces, one above the other. We are looking at Shakespeare's hungry ocean ("When I have seen the hungry ocean gain/Advantage on the kingdom of the shore") in competition with other forces so well matched that it's difficult to declare a winner.

The restless edge of the sea is a wonderful place to witness the passing wasness to isness, as Robert Frost did in "Once by the Pacific."

> The shattered water made a misty din
> Great waves looked over others coming in,
> And thought of doing something to the shore
> That water never did to land before.
> The clouds were low and hairy in the skies,
> Like locks blown forward in the gleam of eyes.
> You could not tell, and yet it looked as if
> The shore was lucky in being backed by cliff,
> The cliff in being backed by continent;
> It looked as if a night of dark intent
> Was coming, and not only a night, an age.
> Someone had better be prepared for rage.
> There would be more than ocean-water broken
> Before God's last Put out the Light was spoken.

SIX

Geology in Ancient Times

There's a legion that never was 'listed,
That carries no colors or crest.
But split in a thousand detachments,
Is breaking the road for the rest."

RUDYARD KIPLING,
"The Lost Legion"

Nicolaus Steno was not the first person to catch a glimpse of the earth's antiquity. There were many before him. The surprising thing about some of these observers was their ability to generalize an understanding of regional geology from one or two seemingly trivial, local situations. We already know Shen Kua, the Chinese philosopher who saw fossil bamboo as evidence of a significant climatic change. That was an impressive feat though not so astounding as his discovery of the origin of the continent of Asia.

This marvelous event took place at the end of the twelfth century A.D. Shen Kua had been traveling again and this time found marine fossils about 330 miles inland from the shore of the Yellow Sea and at the latitude of South Korea.

When I went to Hopei on official duties I saw that in the northern cliffs of the Thai-Hang Shan mountain range, there were

121

belts [strata] containing welklike animals, oyster shells, and stones like the shells of bird's eggs [fossil echinoids]. So this place, though now a thousand li west of the sea, must have once been a shore. Thus what we call the "continent" must have been made of mud and sediment that was once below the water. . . .

Now the Great [Yellow] River, the Chang Shui, the Hu Tho, the Cho Shui, and the Sang Chhein are all muddy, silt-bearing rivers. In the west of Shensi and Shansi the waters run through gorges as deep as a hundred feet. Naturally mud and silt will be carried eastward by these streams year after year, and in this way the substance of the whole continent must have been laid down. *These principles must certainly be true.* (italics added)

Even today, it's almost impossible to think about Shen Kua's untutored insights and not share something of the thrill that he must have felt at the time of first understanding. His last line, "These principles must certainly be true," has the ring of exhultation. Shen Kua had assembled his facts and felt supreme confidence in the least-astonishing judgments that were drawn from them. His genius lay in recognizing the meaning that must be given to data of this quality.

Genius is strange. There's only one way to recognize it and that is after the fact: genius is known only after the completion of a significant accomplishment. Scientific geniuses seem to step out of their own times and into the future. In a very real sense this is true, for their acts create the future. When modern knowledge catches up with them, we see their work as shaping our present level of understanding.

There were other geniuses in China, contemporaries of Shen Kua, who were thinking about geology. The famous neo-Confucian scholar Chu Hsi (1130–1200) gives an equally interesting account of continent building. We gain insight into his mode of thought by comparing several translations of one of his observations. It's important to see how much latitude for interpretation there is in the Chinese calligraphy of those times. Our base will be a literal translation made by Professor Ta-Tseng Ling of the Government Department of Wofford College, Spartanburg, South Carolina: "Once see high mountain have snail shell probably born in rock this rock is old day native snail is water matter low matter but changed high soft matter but changed hard."

Chu Hsi seems to have tried to tell us that he found snail shells in the rocks of high mountains. That much is clear enough, but the second part of the passage seems vague at best until we see these key words interpreted in other translations.

Amadeus W. Grabau (1870–1946), a famous American geologist who worked for the Republic of China from 1920 until his death, gave us this version: "In high mountains there are shells. They probably occur in rocks which were soils of older days, and the shells once lived in the water. Low places became high and the soft mud turned into hard rock."

Professor M. Minato of the Department of Geology and Mineralogy of Hokkaido University, Sapporo, Japan, made another translation for me: "Once I saw spiral and paired (i.e., gastropod and pelecypod) shells in high mountains. The shells were probably present within rock. The rock may have been the soils of older days, the gastropod and pelecypod might be the (sea) water animals. (Accordingly I understood), the low (place) became high (place), and the soft (mud) turned into hard (rock)."

Dr. Joseph Needham's monumental study, entitled *Science and Civilization in China,* contains this account of Chu Hsi's experience: "I have seen on high mountains conchs and oyster shells, often embedded in the rocks. These rocks in ancient times were earth or mud, and the conch and oysters lived in water. Subsequently everything that was at the bottom came to be at the top, and what was originally soft became solid and hard. One should meditate deeply on such matters, for these facts can be verified."

Observations that can be verified are difficult to ignore. They have to be faced as both Chu Hsi and Shen Kua did when they made the same kind of model for the origin of their continent. Similar reactions were undreamed of in Western science until Nicolaus Steno arrived on the scene some five hundred years later. The reasons for this delay in the West may be related to the general intellectual climate of Europe during the Middle Ages and to the limitations imposed by a firm, popular belief in the Biblical flood. Perhaps the real answer is that Chu Hsi and Shen Kua were free to think for themselves and Europeans were not.

Once we begin to recognize the importance of the freedom to think, we're armed intellectually to join Herodotus (484?–425 B.C.) on a field trip in Egypt. Herodotus was a remarkable man who must have been the Lowell Thomas of his day. He seems to have had the knack of getting interesting people to tell him interesting things. All we know about the preparation for his career as a geographer-historian is that he was taught to think in a threefold manner: ask a question, collect information about it, then draw a conclusion to serve as an answer. (This is exactly what we have been doing.) Herodotus chose the whole of the known world as his arena, and what was then known of eternity as the frame for his nine surviving books. We'll examine an excerpt from book 2, romantically entitled *Euterpé, The Muse of Music.* In it he shows himself to be the first regional geologist and scientific oceanographer.

The story is told best in Herodotus' own words. Our presentation of them is taken out of context and rearranged to the extent that extraneous material is deleted. Footnotes are avoided by irreverently interrupting the text to place necessary explanations where they will be most helpful. No attempt has been made to explain Herodotus' references to authorities of his day, or to define his status as a historian. Our main concern is to show how this pioneer was able to draw such magnificent conclusions from the raw field data he found in this region of the world.

> And they [Egyptian priests] told me that the first man who ruled over Egypt was Men, and that in his time all Egypt, except the Thebaic canton [area about 400 miles inland near Thebes (figure 6-1)], was a marsh, none of the land below Lake Moeris [modern Birket Qarum, about 150 miles inland and above Cairo]

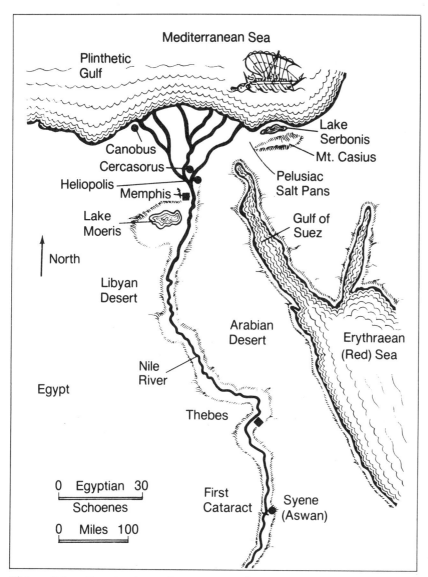

Figure 6-1. Egypt as it was known to Herodotus. (Henry Kiepert, *Atlas Antiquitus*, 1902)

then showing itself above the surface of the water. This is a distance of seven days' sail from the sea up the river. [This much may be thought of as instruction in geology by the Egyptian priests.]

"What they said of their country seemed to me to be very reasonable. For anyone who sees Egypt, without having heard a word about it before, must perceive, *if he has only common powers of observation* (italics mine), that the Egypt to which the Greeks go in their ships is an acquired country, the gift of the river. The same is true of the land above the lake, to that distance of three days' voyage, concerning which the Egyptians say nothing, but is exactly the same kind of country.

The following is the general character of the region. In the first place, on approaching it by sea, *when you are still a day's sail from the land, if you let down a sounding-line you will bring up mud, and find yourself in eleven fathoms [of] water, which shows that the soil washed down by the stream extends to the distance.* (italics mine)

Herodotus must have been curious enough to do it. The passage also implies that he had a general knowledge of the sea floor and thought that unsorted, clay-rich river muds were distinctly unusual.

From the coast inland as far as Heliopolis [very close to modern Cairo and at the head of the distributaries of the Nile], the breadth of Egypt is considerable, the country is flat, without springs, and full of swamps. As one proceeds beyond Heliopolis up the country, Egypt becomes narrow, the Arabian range of hills which has a direction from north to south, shutting it in upon the one side, and the Lybian range on the other. The former ridge runs on without a break, and stretches away to the sea called the Erythraean [Red Sea]; it contains the quarries whence the stone was cut for the pyramids of Memphis [Cheops and others]: and this is the point where it ceases its first direction, and bends away in the manner above indicated.

The greater portion of the country above described seemed to me to be, as the priests declared, a tract gained by the inhabitants. For the whole region above Memphis, lying between the two ranges of hills that have been spoken of, appeared evidently to have formed at one time a gulf of the sea. It resembles, to compare small things with great, the parts about Ilium and Teuthrania, Ephesus, and the plain of the Maeander. In all of these regions the land has been formed by rivers, whereof the greatest is not to be compared to any of the five mouths of the Nile. I could mention other rivers also, far inferior to the Nile in magnitude, that have effected great changes. Among these not the least is the Achelous, which after passing through Acarnania, empties itself into the sea opposite the islands Echinades, and has already joined one-half of them to the continent.

The range of field examples available to Herodotus is extremely interesting. He was familiar enough with the general character of river action to cite two examples in Greece, some two hundred miles northwest of Athens, and three examples in modern Turkey, about two hundred miles east of Athens (figure 6-2). These features are similar to those of the Nile six hundred miles to the southeast.

The Ilium area (figure 6-3*A*) is on the west coast of Greece about ten miles from the Albanian border. The flat depositional plain of the Thiamus River is readily distinguishable from the surrounding hills. The long point is the delta of the river as it is being built into the Ionian Sea.

Teuthrania is on the mainland of modern Turkey (figure 6-3*B*) just east of the island of Lesbos. The delta that is being built on the Aegean shelf has been modified by wave action, which has produced the rounded form.

The Ephesus area (figure 6-3*C*) is also on the coast of Turkey, just northeast of the island of Samos. The Kücük River flows across

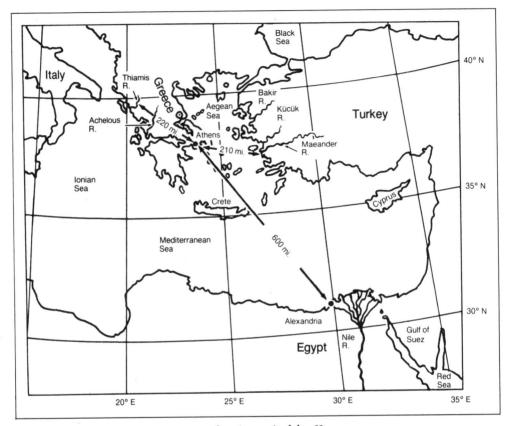

Figure 6-2. Here are the rivers cited by Herodotus as field evidence to support his geologic conclusions.

a narrow depositional plain into the Aegean Sea. The Maeander River flows across a wide depositional plain between steep, hilly walls. The winding contortions of the Maeander are easily seen on the map. Our use of the term *meandering stream* is derived from the name and nature of this river.

The Achelous River (figure 6-3D) is in central Greece at the mouth of the Gulf of Patras, where it opens into the Ionian Sea. The delta of the Achelous obviously has been built out from the land and

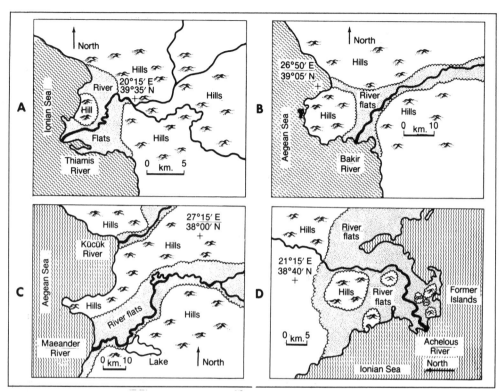

Figure 6-3. These maps were redrawn from a series used by the German army in World War II. The details known to Herodotus are fully evident today. He compared the work of these rivers (in what are now Greece and Turkey) to the work of the Nile.

has connected the series of isolated islands, so that they now appear to be hills on the delta.

One of my wise teachers insisted that geology is an outdoor science that could be learned in only one place, the field. This aspect of the science does not change. Herodotus had become familiar with the geology of Greece and Turkey before he saw that of Egypt. His amazing ability to generalize about the significance of similarities among river plains and deltas is the key to his competence as a field geologist.

> In Arabia, not far from Egypt, there is a long and narrow gulf running inland from the sea called the Erythrean, of which I will here set down the dimensions. Starting from its innermost recess and using a rowboat, you take forty days to reach the open main while you may cross the gulf at its widest part in the space of half a day. In this sea there is an ebb and flow of the tide every day.

At this point it is very important to determine the exact place (figure 6-1) Herodotus had in mind. The data are a little contradictory. *Webster's Third International Dictionary* assures us that the Erythrean Sea is the Red Sea, citing the Greek root word, *eruth* for red as evidence. Other authorities differ and identify it as the Arabian Sea but that does not fit the description. The real problem concerns the dimensions. The Red Sea is between 100 and 200 miles wide and 1,200 to 1,300 miles long. No one could row across in half a day. The Gulf of Suez is only 30 miles wide, but it is 180 to 200 miles long. We shall assume that Herodotus was confused by hearsay evidence and meant the Gulf of Suez and some part of the northern end of the Red Sea in his next comments. This is an attractive idea, for the dimensions of the landforms are quite similar to those of the Nile valley and the more open delta country he has already described. The comparison of the Nile valley and the Gulf of Suez–Red Sea area, as shown in figure 6-1 is quite apt.

> My opinion is, that Egypt was formerly very much such a gulf as this—one gulf penetrated from the sea that washes Egypt on the north and extended itself towards Ethiopia; another entered from the southern ocean, and stretched towards Syria; the two gulfs ran into the land so as to almost meet each other, and left between them only a very narrow tract of country. Now if the Nile should choose to divert his waters from their present bed

into this Arabian gulf, what is there to hinder it from being filled up by the stream within, at the utmost, twenty thousand years? For my part, I think it would be filled in half the time. How then should not a gulf, even of much greater size, have been filled up in the ages that passed before I was born, by a river that is at once so large and so given to working changes?

In a logical sense Herodotus is guilty of begging the question. Nevertheless his grasp of the fact of change through time and of the reality of long spans of past time is truly remarkable. The Nile valley has been cut into the higher plateaus on either side by the erosion of the river during periods of much lower sea level. In the glacial ages that began about 2 million years ago and ended less than 20,000 years ago, the sea level was lowered by several hundred feet as water was accumulated in the extensive ice sheets. All of the major coastal rivers were able to cut more deeply at that time. Most of the fine harbors of the world are flooded river valleys. Actually two causes work together to account for them.

Flooding represents a high sea level due to the melting of Pleistocene ice sheets. Harbors rather than deltas at the coastline are evidence that the rivers do not supply enough sediment to fill the old valleys. Conversely, deltas rather than harbors at river mouths indicate excessively large sediment loads and rapid deposition. Both the Nile and Mississippi Rivers have large deltas and sediment-filled valleys for hundreds of miles upstream. Much of this material was deposited since glacial times. On the other hand, the Hudson has not filled New York harbor since the flooding. Neither have the San Joaquin or Sacramento Rivers filled San Francisco Bay. Chesapeake Bay is the flooded mouth of the Susquehanna River. Hampton Roads is a part of the Chesapeake Bay and also the flooded mouth of the James River. The first ironclads, the *Monitor* and the *Merrimac,* battled here during the hostilities of 1862. So we see that Herodotus was not really very far off in his estimate of the amount of time that large, silt-laden rivers require to fill major estuaries and valleys.

The Gulf of Aden, Red Sea, Gulf of Suez, Gulf of Aqaba, and the lowland trench of the Holy Land are remarkable, troughlike structures bounded by steep walls of broken rock. These breaks or faults, were formed as Africa and Arabia moved apart in the process of continental drift. Frequent earthquakes that rock the southern part of the Red Sea today are testimony that the process is still in action. The floors of all the troughlike structures have been dropped below sea level in much the same manner as a roof would fall if the walls of a building were moved apart. The Sea of Galilee, Jordan River, and Dead Sea (the latter with a surface 1,286 feet below sea level) are all bodies of water confined in the same trough that holds the Gulf of Aqaba. Herodotus compared the external aspects of the lowland regions of the Nile Valley and Gulf of Suez, not their internal aspects. His real intellectual breakthrough consisted in bringing a regional situation into focus and considering change through time in a rational manner.

> Thus I give credit to those from whom I received this account of Egypt, and am myself, moreover, strongly of the same opinion, since I remarked that the country projects into the sea further than the neighboring shores, and I observed there were shells on the hills and that salt extruded from the soil to such an extent as even to injure the pyramids.

These famous lines are often quoted as proof that Herodotus was the first paleontologist. He saw shells in the rocks and he called them shells. This is a bold, straightforward approach. Herodotus' lines are also startling in their recognition that, because the Nile Delta protrudes into the Mediterranean, the bulge is a record of the extension of the Nile delta by deposition.

If Herodotus had sailed to Egypt by navigating the eastern coastal waters of the Mediterranean, he would have had to have changed course to pass around the curve of the Nile delta before he could have reached the western distributary mouth near modern Alexandria. We do not know where he entered Egypt, but obviously he was aware of the delta form. The average tourist would probably not consider that the shape of a shoreline had special meaning, but Herodotus already had seen the same effect in the delta of the Achelous River, which protruded and tied half of the Echinide Islands to the mainland (figure 6-3*D*). The delta of the Thiamis River (figure 6-3*A*) protrudes in a similar way. These are further examples of the manner in which Herodotus used specific observations to draw generalized conclusions.

> I noticed that there was but a single hill in all of Egypt [speaking of the confined Nile valley] where sand is found, namely the hill above Memphis; and further, I found the country to bear no resemblance either to its borderland Arabia, or to Lybia—nay, nor even to Syria, which forms the seaboard of Arabia; but whereas the soil of Lybia is, we know, sandy and of a reddish hue, and that of Arabia and Syria inclines to stone and clay, Egypt has a soil that is black and crumbly, as being alluvial and formed of the deposits brought down by the river from Ethiopia.
>
> One fact which I learnt of the priests is to me strong evidence of the origin of the country. They said that when Moeris was king, the Nile overflowed all Egypt below Memphis, as soon as it rose so little as eight cubits [12 feet]. Now Moeris had not been dead 900 years at the time when I heard this of the priests; yet at the present day, unless the river rises sixteen, or, at least, fifteen cubits [22.5 to 24 feet], it does not overflow the lands. It seems to me, therefore, that if the land goes on rising and growing at this rate, the Egyptians who dwell below Lake Moeris, in the delta, as it is called, and elsewhere, will one day, by stoppage of the inundations, suffer permanently the fate which they told me they expected would some time or other befall the Greeks [famine by drought].

Herodotus speaks of time with the ready familiarity of a scholar who has lived with the wasness of the is. His reference to an event nine hundred years past is comparable to one of our contemporaries saying, "The Norman conquest had taken place only nine hundred years before I heard about it." Herodotus's discussion of the rate of deposition is difficult to evaluate. The Egyptians did have Nilo-meters to measure the rise and fall of the river. Some of them were more than one thousand years old in Herodotus's time, so that the priests could have had valid measurements on which to base their information. At the least his statements are an acknowledgement that past change is a reality and that further change may be anticipated.

> If then we choose to adopt the views of the Ionians concerning Egypt, we must come to the conclusion that the Egyptians had formerly no country at all. For the Ionians say that nothing is really Egypt [but] the Delta, which extends along the shore from the Watch-tower of Perseus, as it is called, to the Pelusaic Salt-pans [just east of modern Port Said], a distance of forty schoenes [about three hundred miles by Herodotus's conversion factors], and stretches inland as far as the city of Cercasorus [near modern Cairo], where the Nile divides into two streams which reach the sea at Pelusium and Canobus [near modern Alexandria] respectively. The rest of what is accounted Egypt belongs, they say, either to Arabia or Lybia. But the Delta, as the Egyptians affirm, and as I myself am persuaded, is formed of the deposits of the river, *and has only recently, if I may use the expression, come to light.* If, then, they had formerly no territory at all, how came they to be so extravagant as to fancy themselves the most ancient race in the world?. . . . But in truth I do not believe that the Egyptians came into being at the same time with the Delta, as the Ionians call it; I think they have always existed ever since the human race began; as the land went on increasing, part of the population came down into the new country, part remained in their old settlements. . . . (italics added)

It's interesting to abstract from these writings the underlying principles that shaped Herodotus's perceptions and observations.

Think of the following list of principles as we did those of Steno, as a set of keys that anyone can use to unlock the history of the earth. These may be called "the first, first principles of geology." Herodotus must have known them in order to have been able to have done what he did.

1. He acted on the basic presupposition of science: nature can be understood.

The prima facie case for this is the way he comprehended all he saw.

2. He collected and classified field data with great care.

An example of this method is his observation on the composition of the mud floor of the Mediterranean Sea. Another is his citation of the types of soils and rocks associated with different regions.

3. He searched for the general, or universal, natural characteristics that relate classes of phenomena; he was not content to accept a unique explanation for each separate item.

This is one of the most important aspects of his work. The average person would not think of comparing the valley of the Nile with the appearance of river valleys hundreds of miles away and far less spectacular in size.

4. He grasped the wasness of the is.

It had become possible to view the pyramids, already ancient by the time Herodotus arrived, as monuments not only to monarchs but also to the time elapsed since their construction. Nevertheless, it is remarkable that Herodotus was able to adopt the same perspective with regard to a natural phenomenon like a river delta, seeing it as a monument to a geologic action occurring through time (much as we observed the growth of the block of redwood in chapter 5).

5. He understood that causality could be comprehended through a least-astonishment approach.

He used the standard format of a list of facts that implied a conclusion. He singled out anomalies, restructuring his conclusions, rather than accepting the conclusions of others. His discussion of the northward movement of the Egyptian people onto the delta as it was increased in size by deposition is a good example of his use of the least-astonishment approach.

Taken together, these five generalizations constitute the first, first principles of geology. What would have to be added so that these principles would be applicable to modern science? Herodotus didn't know the chemistry and physics necessary to study materials, and his classification procedures were limited. He didn't know the full range of processes by which the uniformitarian analogy is used in least-astonishment solutions. He didn't know the biology necessary to make the most of fossil evidence. Base maps for displaying information in picture form had not been developed, so that his ability to compare regional data was more limited than his skill at making observations. Despite these difficulties, the insight Herodotus displayed may be explained by a single word: genius.

NEAR EASTERN GEOLOGIC IMAGERY

The history of science is an amazing study of false leads, blind alleys, dead ends, and all manner of intriguing insights. The pattern must be unraveled either on a de facto basis, as in the case of Stonehenge, or with the surviving literature, whatever it may be. If we look back to the time before Herodotus to see what might have been known about geology, there are scattered bits of information but no comprehensive description of a region. Literature is a useful source of information because whatever the time or place, it reveals the concepts generally understood by its audience. An author who used a geologic image to emphasize a point must have felt that the image would be understood. We shall examine an example of such usage in the Old Testament. The setting is familiar enough. About 539 B.C. Cyrus permitted the Jews to return from Babylon to Jerusalem across the arid Arabian Peninsula.

A voice cries:
In the wilderness prepare the way of the Lord,
Make straight in the desert a highway for our God (Isaiah 40:3).
Every valley shall be lifted up,
and every mountain and hill shall be made low;
the uneven ground shall become level,
and the rough places a plain (Isaiah 40:4).

This geologic imagery may not seem impressive to those of us raised in humid regions where rivers flow to the sea, disposing of the sediment in an out-of-sight, out-of-mind manner. However, Isaiah and his contemporaries, to whom his words had rational as well as spiritual meaning, were natives of an arid region with few permanent streams and totally different geologic actions.

American geologists became fully aware of these climatic effects when they began to study the Basin and Range Topographic Province. This is an enormous, irregularly shaped desert region that extends from southern Oregon to Mexico City and lies behind the protection of the Cascade, Sierra Nevada, and Sierra Madre Occidental ranges. The system of erosion, transportation, and deposition of rock fragments from the crests of the mountains to the broad valleys would be more familiar to Isaiah and the Jews of Babylon than to residents of our own eastern seaboard. Figures 6-4 and 6-5 illustrate how mountains in this province are eroded and buried in their own debris. Ideally, the end product of this kind of action should be the development of a nearly flat plain, just like the one Isaiah described. Under these conditions, every valley would be filled as every mountain and hill is worn away. The desert climate is so dry that there isn't enough water flowing to support rivers that can transport rock fragments to the sea. All of this is seen in the illustrations.

The first time I flew west across this area, its resemblance to a battlefield astounded me. Fault-block mountains are still rising and competing with the destructive forces of erosion. Occasional flash floods strip sands and gravels from the knife-sharp mountain ridges and wash them down to the flat floors of the valleys. Steeply sloping

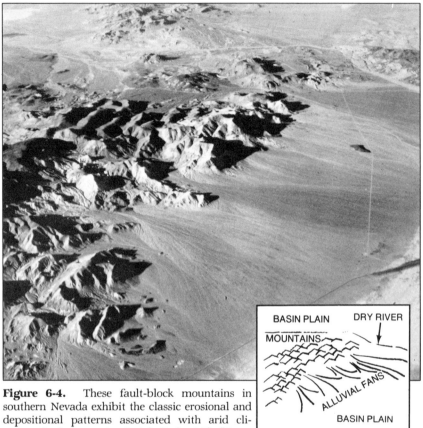

Figure 6-4.   These fault-block mountains in southern Nevada exhibit the classic erosional and depositional patterns associated with arid climates. St. Luke quotes Isaiah (Luke 3:5): "Every valley shall be filled,/Every mountain and hill shall be brought low."

aprons of alluvial fans may be seen separating the mountain crests from the depositional plains. Bit by bit the valleys become filled and their floor levels rise toward the tops of the now-subdued mountain peaks.

Most of the runoff water either evaporates or sinks into the porous sands. Occasionally, after very heavy rains, temporary rivers are able to form and drain away for short distances to reach and help fill nearby deeper valleys. An example of a riverbed of this type is partially blocked from view by the jet engine in figure 6-4. A hint of Isaiah's final development stage is shown in figure 6-5. The Biblical reference to this kind of geology is so clear that it's hard to imagine another two thousand five hundred years passed before anyone formalized it into a branch of science.

All of this surged through my mind in the late 1960s when I first discovered that Isaiah and St. Luke were geologists. My next step was to check the translations of critical Hebrew words used in the text. I thought there might be more geology hidden in these verses than previous translators had appreciated. Sure enough, the verb

Figure 6-5.   We are looking into Mexico from a point about one hundred miles west of El Paso, Texas. Arid-style erosion and deposition are well advanced here as they were west of Babylon in the time of Isaiah (Isaiah 40:4): "The uneven ground shall become level,/and the rough places a plain."

form *made low* at the end of the second line in Isaiah (40:4) is the Hebrew word *shaphel*. One of the Biblical scholars who was helping me said, "You know, that's odd! The pre-Mosaic name for the area between the Palestinian coastal plain and the highlands flanking the Dead Sea some distance northeast of Gaza is Shaphelah, or 'place made low.' "

This is probably not a coincidence. The ancient Palestinians must have recognized the fact that erosion had lowered this area of relatively soft chalks, cherty limestones, and limey clays. Ideas of this sort may seem impossibly strange to persons living in a humid area where thick mats of vegetation conceal and protect the rocks. If so, a glance at figure 6-6 may change their minds.

This starkly eroded mountainside is in the desert of Afghanistan. Raw edges of the beds reveal the parts of the rocks that have

Figure 6-6. The folded rocks and eroded mountain slopes of arid Afghanistan reveal a story that is easily read. (Courtesy, Charles Luke Powell, Jr.)

been eroded. Look at the sloping platform just to the right of the massive rock of the basalt plug. It would be hard to avoid the fact that this is a shafelah, a place made low. Modern geologists have another name for it—*pediment*—taken from the vague resemblance to the triangular geometry of a classical Greek temple. The ancient wisdom of using the word shafelah seems quite appropriate.

One of the most startling proofs of ancient geological knowledge is to be found in the design of the water mines of the Middle East. The classic type, called a *qanát,* is used to take water from the interior of an alluvial fan and make it available for irrigation in the valley below. The mine consists of a line of vertical shafts dug into a fan in order to intersect water-bearing gravel beds. The shafts are all connected by a tunnel dug on the slope of the beds from an outcropping position up toward the head of the fan. The shafts and tunnels are lined with porous terra-cotta pipe both to prevent collapse and to permit water to enter. In Iran alone there are over 22,000 of these qanats with a total length of about 175,000 miles. The average length of each mine is approximately 8 miles. Many of them have been maintained in use since ancient times.

Imagine the task of digging a qanat far underground through dangerous slurries of wet sand and gravel by the wavering light of a primitive oil lamp. Each advance would require immediate structural support by fitting curved sections of terra-cotta pipe into place before the roof and walls collapsed. Now imagine engineering skills so well developed that the total tunnel length is as great as seven trips around the world. The engineers who did that understood geology.

No discussion of the science of geology in ancient times would be complete without acknowledging the work of two Greek geographers, Pytheas (circa 320 B.C.) and Eratosthenes (257–180 B.C.). They measured the shape and size of the earth. The story begins sometime in the late fourth century B.C. when Pytheas sailed north from the Strait of Gibraltar on a voyage of discovery. His course led him through the Irish Sea. Pytheas continued to a place he called Thule (perhaps Iceland or some part of Norway), where "the Sun sleeps only a few hours below the horizon." Although poorly remembered today, this voyage ranks with those of Christopher Columbus and Captain James Cook, for Pytheas added a new dimension to our knowledge of the planet.

People had suspected for some time that the earth must be curved because ships disappeared, hull first, over the horizon and the earth cast a curved shadow on the moon every lunar month. Yet no one had approached the question as directly as Pytheas did. His method was based on simple geometry. If the earth was flat, the vertical angle betweeen the horizon and the sun at noon would be constant no matter where it was measured. Pytheas sailed north to test this hypothesis and found that the sun angle became systematically smaller the farther he went. The earth was a ball!

Eratosthenes, head librarian at the great intellectual center in Alexandria, was a man of unusual imagination. He realized that it was possible to measure the size of the earth by simultaneously measuring two noontime sun angles at opposite ends of a north-south line of known length. The method Eratosthenes used in 215 B.C. is shown in figure 6-7.

Eratosthenes knew that there was a deep water well near Syene at the first cataract of the Nile, some five thousand stades (a unit of distance) south and slightly east of Alexandria (see figure 6-7A). This water well was almost exactly on what is now called the Tropic of Cancer. At noon on the longest day of every year the sun is directly overhead and, in Eratosthenes's time, shone right into the well and was reflected back from the water's surface. Eratosthenes realized that at this precise moment, a straight line would connect the sun, the well, and the center of the earth, as shown in figure 6-7B. What an opportunity! It was at this moment in Alexandria, therefore, that Eratosthenes set up a device, called a scaphe (figure 6-7C), to measure the angle between the parallel rays of the sun and a vertical rod pointing towards the center of the earth. (A scaphe is a hemisperical bowl, calibrated to divide a quarter circle into fifteen equal units of four degrees each; a thin rod, attached to the bottom of the bowl, extends upward from its exact center to the height of the rim.) Eratosthenes placed the scaphe on the ground in a sunlit spot and

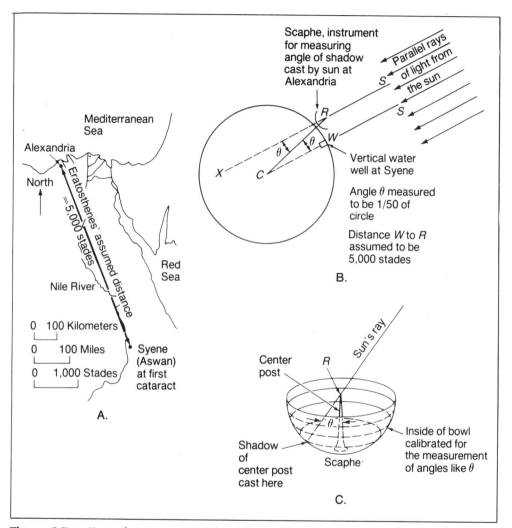

Scaphe, instrument for measuring angle of shadow cast by sun at Alexandria

Parallel rays of light from the sun

Mediterranean Sea

Alexandria

North

Eratosthenes' assumed distance = 5,000 stades

Nile River

Red Sea

0 100 Kilometers

0 100 Miles

0 1,000 Stades

Syene (Aswan) at first cataract

A.

Vertical water well at Syene

Angle θ measured to be 1/50 of circle

Distance W to R assumed to be 5,000 stades

B.

Sun's ray

Center post

R

Shadow of center post cast here

Scaphe

Inside of bowl calibrated for the measurement of angles like θ

C.

Figure 6-7. Eratosthenes measured the polar circumference of the earth in 215 B.C. This was his instrumental layout.

filled it with water, adjusting the position of the bowl until he knew the rim was horizontal and the rod vertical. He drained the water so that the tip of the rod R could cast a shadow on the inside of the bowl. From this he could read the value of angle θ, one-fiftieth of a full circle, or in modern terms, seven degrees and twelve minutes. That bit of news doesn't seem to be very important until we look back at figure 6-7B and see the geometry represented by the lines and angles.

One of the theorems of plane geometry states that when parallel lines are cut by a diagonal straight line, the alternate interior angles are equal. Eratosthenes had measured the angle at the center of the earth between the line of the sun shining down into the well at Syene and the line of the vertical rod in the scaphe. That value was one-fiftieth of a full circle. The arc length on the surface of the earth represented by this angle was five thousand stades, so the circumference of the earth must be fifty times five thousand, or two hundred and fifty thousand stades. All we need to know now is the length of a stade, and we can check Eratosthenes' measurement against the modern value, 24,900 miles.

Unfortunately, there are several possibilities, and we don't know which to choose. A Roman stade was the length of a stadium, 607 feet. In this case, his measurement was too large by a factor of 15.6 percent. An Athenian stade was 582.7 feet, resulting in a calculation of the earth's circumference that was 10.8 percent too large. The Egyptian stade of that time was equal to 516.7 feet, giving a measurement that was only 1.6 percent too small. All of these figures are interesting but not of critical importance.

The significant point is that Eratosthenes devised a way to prove the earth to be of finite size and considerably larger than the part of it that was known to the Greeks. With that start we might imagine that exploration would progress rapidly. Obviously, that didn't happen for a number of reasons. Some of them offer significant insights for understanding the relationships among mankind, science, and society.

Eratosthenes had no Henry Oldenburg to translate his ideas into readable form and advertise them widely among the scientific community. There were no scientific journals to be read by the intellectuals of the day. Caesar's troops, battling the Egyptian population of Alexandria in 48 B.C., burned a large part of the great library and destroyed about five hundred thousand volumes of the precious lore of antiquity. Political leaders are power brokers more often concerned with local issues and conquest than they are with allocating funds to explore the nature of nature. In short, very few people really cared about the size and shape of the earth. There is a parallel here with the fate of our modern, government-sponsored space programs. In a very real way we are still living in the time of Eratosthenes and have seen, but failed to understand, his smile of satisfaction on that long day in 215 B.C.

Christopher Columbus was one of the first people to use the ideas of Eratosthenes in an imaginative way, but his voyages are rarely seen in that context. The story that is usually told focuses on the man rather than on the earth. In the late summer of 1492 Columbus sailed west with a ragtag crew, said to be ready to mutiny for fear of falling off the edge of the earth. However, Columbus was staunch and commanded them to "Sail on!" At last land was sighted, and the mutterings ceased. A landing was made on San Salvador, a small island at the eastern edge of the Bahamas. Columbus thought he had found India and named the inhabitants, Indians. He spent some time sailing around looking at other islands and then turned back to Europe to spread the news that he had reached India by sailing west. Everyone was pleased.

This silly, little view of history places too much emphasis on men floundering in ignorance and not enough on the real world in which they lived. Columbus and crew were actually caught in the moving parts of a great machine. It picked them up in Europe and delivered them, by the power of the sun's heat and the earth's gravity, to San Salvador. (Thor Heyerdahl used the same machine when he sailed the *Ra I* and *Ra II* from Africa to America nearly five centuries later.)

Sunlight strikes the earth more directly in the equatorial zone than it does at the poles. This makes tropical air hotter and therefore less dense than polar air. The effect of gravity on these air masses is to drive polar air along the ground toward the equator. At the same time, tropical air is lifted to heights of about ten miles where it flows back toward the poles and maintains the circulation. Air flow is complicated by a number of factors, however, including the changes of the seasons, the rotation of the earth, and the limited space near the poles. The equatorial zone is close to 24,900 miles in circumference. Not all of the air that rises in this belt can pass down again through the two small polar areas. As a result, there are additional places, called standing highs, where cool air falls back to the earth's surface with a reasonably uniform rotating motion. The trade winds and great patterns of ocean circulation, moving slowly around and around, are the result of this falling and spinning action. Columbus took advantage of one of these systems to sail to America and return to Europe.

Leaving Palos, Spain, on the third of August, 1492, he sailed downwind to the Canary Islands off the coast of northwest Africa, as shown in figure 6-8A. The *Pinta* was lateen-rigged in the manner of many ships that operated in the Mediterranean Sea. It couldn't keep up with the more efficient square-rigged *Nina* and *Santa Maria*. The expedition halted in the Canaries long enough to rerig the *Pinta* and then continued sailing westward and downwind at speeds between three and five miles per hour. On the twelfth of October it reached San Salvador.

It is probable that the potential mutiny of Columbus's sailors was due more to a fear of running out of food and water than to falling off the edge of the earth. Columbus certainly knew how difficult it would be to try to tack his ships back against the combined forces of wind and current, but he probably withheld his plan to return by sailing north to the latitude of Portugal and using a known tailwind that blew from west to east during the winter months. The return voyage was begun from Hispaniola, a large island between Cuba and Puerto Rico, on January 4, 1493. The course lay northeast, tacking into the dominant winds until the latitude of

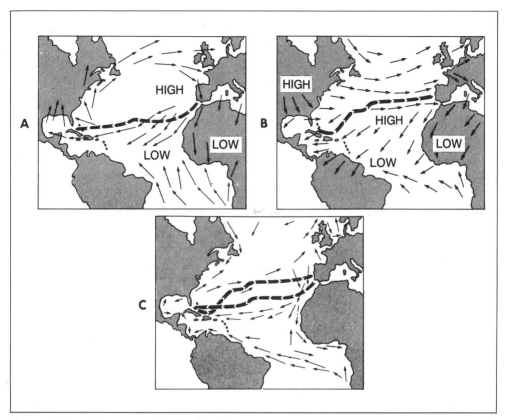

Figure 6-8. The voyage of Columbus in 1492 was controlled to a great extent by the prevailing winds and ocean currents! *A)* summer winds, *B)* winter winds, and *C)* ocean currents.

Portugal was reached, near the fortieth parallel. From here the course was due east with the winter trade winds behind them, as shown on figure 6-8*B*. A landfall was eventually made at Lisbon on March 15. The full trip, over and back, extended the static earth model of Eratosthenes to include a new view of flowing surface features.

Columbus's model of sea and air motion as a means of moving

ships back and forth between Europe and America was pragmatic rather than scientific. Scientists didn't pay much attention to it, but sea captains and hardware collectors did. Cortez, Balboa, Coronado, Ponce de Leon, de Soto, Raleigh, Smith, Drake, Cartier, and Hudson were all in the hardware business; they liked the smell of gold. The scientists came later.

I was on San Salvador in March 1956. A company of Army Engineers was there to build a hush-hush tracking station in preparation for the coming space age. They gave me a standing ovation as I climbed out of our Grumman Goose. It was Saturday noon, and the soldiers thought I was a liquor salesman. We all got along quite well. I have often wondered what the same men might have thought of Columbus's crew barging in unannounced like that. Five hundred years makes a great deal of difference.

Matthew Fontaine Maury (1806–1873) was a naval officer, crippled in a stagecoach accident in 1839 and forced to develop a new career. Maury found his destiny in 1842 when he was appointed to supervise the Navy's Depot of Charts and Instruments. In 1855 he published a classic book, entitled *Physical Geography of the Sea and Its Meteorology*. The preface is self-explanatory:

> The primary object of *The Wind and Current Charts*, out of which has grown this treatise . . . was to collect the experience of every navigator as to the winds and currents of the ocean, to discuss his observations upon them, and then to present the world with the results on charts for the improvement of commerce and navigation.
>
> By putting down on a chart the tracks of many vessels on the same voyage, but at different times, in different years, and during all seasons, and by projecting along each track the winds and currents daily encountered, it was plain that navigators hereafter, by consulting this chart, would have for their guide the results of the combined experience of all those whose tracks were thus pointed out. . . .
>
> Such a chart . . . would show [a young navigator] not only

the tracks of [a thousand vessels] . . . but the experience also of each master as to the winds and currents by the way, the temperatures of the ocean, and the variation of the needle. All of this could be taken in at a glance, and thus the young mariner, instead of groping his way along until the lights of experience should come to him by the slow teachings of the dearest of all schools, would find here at once that he had already the experience of a thousand navigators to guide him . . . as though he . . . had already been that way a thousand times before.

We have come a long way from the descriptions of earth science by Isaiah. Here at last, in 1855 A.D., we find a man who knows that his duty is to communicate technical information so that someone else can use it as technical information. The qanat engineers would have appreciated that, for their lives were at stake if things went wrong.

We've seen Kipling's Lost Legion pass by

The qanat diggers,
Herodotus and Columbus,
Chu Hsi and Matt Maury,
Eratosthenes and Pytheas,
Shen Kua and Second Isaiah.
They broke the trails that led us here.
Fare them well.

SEVEN

Adventures with Unconformities

> "In that first moment of shock, with my mind already exploding beyond old boundaries, I knew that something had happened to the way I looked at things."
>
> COLIN FLETCHER
> *The Man Who Walked Through Time*

Rock-shock" is a very real experience. It is a state of surprise that often sweeps unexpectedly over a geologist lost in the quiet company of rocks. My own first experience of rock-shock began with a chance conversation. A friend told me where to find pebbles of nearly 600 million-year-old Erwin quartzite in a railroad cut about a mile north of the James River near Richmond, Virginia. These pebbles were of special interest because they contained fossil worm borings called *Scolithus linearis* (literally, "worm-rock linear"). I knew two things about them that lent excitement to the hunt. Worms are our ancestors in the evolutionary chain of development. We share several aspects of their anatomy: red blood, a central food canal from mouth to anus, a linear nerve chord, and body symmetry. Finding ancestral roots is always fun. However, a much more important reason for looking for these pebbles was to verify their appealingly romantic history.

The source rocks lay on the western flank of the Blue Ridge Mountains, about 150 miles up the James River from Richmond. I had been on the outcrop and seen the worm tubes; so I knew what to look for. All I wanted to know was whether or not these quartzites were actually hard enough and tough enough to withstand the beating they would have been subjected to while being rolled that far by the river. Pebbles as large as Idaho potatoes require flood water velocities on the order of 20 miles per hour to pick them up and move them along. Water moving this fast has tremendous force and would seem to be capable of doing a great deal of damage to even the toughest rocks far upstream long before depositing them at the edge of the coastal plain. I hurried on to the outcrop, bursting with anticipation.

Everything was just as I had been told it would be. The quartzite pebbles were there as large as Idaho potatoes, and the worm tubes could be identified with ease. All I had to do was reach up shoulder high, above a massive, partially weathered, gray granite, and dig out as many pebbles as I wanted. It was all so easy. I had leaned my body against the granite for balance and begun to dig a pebble out of the soft brown matrix of the coastal plain conglomerate when my attention was distracted by a sharp object cutting into my kneecap. I looked down and saw that I was leaning against the jagged edge of a quartz vein. My eyes followed the line of the vein upward to the top of the granite in a most routine, disinterested way. A cow could have been no more intellectually placid than I was at that moment. But then I began to think as a geologist should.

The quartz vein stopped at the top of the granite without penetrating the soft, porous conglomerate above it. I remembered how quartz veins are formed: by mineral-bearing, hot water under high pressure, capable of intruding the rock around it. Quartz is precipitated as a solid whenever the hot water cools enough to lose its ability to keep the mineral in solution. Ah hah! I suddenly knew why the quartz vein stopped at the top of the granite. The conglomerate was not present when the quartz-bearing waters were moving up. That made sense. However, granite is an igneous rock, a liquid, too, in origin. There was no sign of granitic material intruding into the conglomerate either. Therefore, the granite was also older than the conglomerate. These rocks were of different ages. They belonged in different time-bags. I was on the track at last. Here was an anomaly, and I was beginning to think about it as a scientist should.

My first conclusion was that the top of the granite had been an ancient erosion surface, a part of the landscape on which the sun once shown, as it does today on our own surface rocks. That was easy to prove. I could see a few granite pebbles mixed in among the other rocks in the capping conglomerate. The granite pebbles were comparable in every way with the loose rocks weathered out and tucked back into the crannies of modern outcrops. I was elated but still unaware of the staggering rock-shock that was soon to strike.

My attention flickered to an image that appeared as clearly to me as a well-focused photograph. It was an image, or better, a

Figure 7-1.   This is the way I remember my rock shock at seeing my first unconformity between early Carboniferous granite and mid-Tertiary Coastal Plain conglomerate.

model of the original granite mass as it must have extended for dozens of miles to the north and south. I realized that it would take a great deal of vertical erosion to expose all of this granite. The restricted valley of the James River was not that wide. I was looking at a big event! I could take that. Geology is full of big events. The rock-shock that struck without warning came when I realized that I was actually seeing something that wasn't there: eroded ghost rocks. One minute I was a self-assured and reasonably stoical 22-year-old mining engineer, hammering happily on a rock. Thirty seconds later I was totally out of control, sitting on a railroad track sobbing in unrestrained creative joy. I had had a staggering thought. In Colin Fletcher's words, ". . . something had happened to the way I looked at things!"

Imagine finding a flat and apparently empty time-bag with about 310 million years in it, but nothing else. The time-bag below was properly filled with a 330 million-year-old early Carboniferous granite. Just above the empty time-bag was another one containing a 20 million-year-old mid-Miocene conglomerate. Yet right there, in the middle, on a paper-thin line, separating granite from conglomerate, was this seemingly illegitimate time-bag with absolutely nothing in it.

I looked up at the sky and filled my mind's eye with a new image: a model of the way things looked back in liquid-granite time. Molten granite was present at the level of the railroad track. This hot mass extended upward for several miles to the place where intrusion of the overlying sediments was still going on, exactly as we have seen it with Dr. Hutton in Glen Tilt in Scotland. This image included an additional ten vertical miles of sediment before reaching daylight. I imagined a sun-drenched surface covered with late Paleozoic plants growing in the shadow of an erupting volcano at least another two miles higher.

When my mind returned again to 1940 A.D. and I looked at the exposed rocks in the railroad cut, I was overpowered by the knowledge of what I had seen. Twelve or more vertical miles of material had been stripped away and sent to the sea by rivers that once flowed across this area between liquid-granite time and conglomerate time. No wonder the middle time-bag had no rocks in it. It was full of 310 million years of erosion! Rock-shock struck again. I knew I had seen twelve vertical miles of erosion with my own eyes. Think of twelve sets of Grand Canyons stacked one above the other and so completely eroded that the wall rocks are also missing. The idea was stupendous; it was the first *unconformity* I had ever appreciated.

An unconformity is a buried surface of erosion separating younger rocks on top from older rocks below. Part of the exciting adventure of recognizing unconformities is in knowing that these things were once landscapes. They have been exposed to the warmth of the sun, to the chill of snow, to the pelting of rain, and to capricious winds. These ancient landscapes have been covered with plants, traversed by animals, insects, pteradons, and birds. They are geography books that can be used to define the changing positions of land and sea through time. Unconformities are fun because they are windows through which we can see so many wonderful things.

Grove Karl Gilbert, chief geologist with the United States Geological Survey, wrote of his encounter with an unconformity while traveling in 1875 with pack mules from Salt Lake City to the Henry Mountains in Utah:

> No one but a geologist will ever profitably seek out the Henry Mountains, and I will therefore, in making out a route by which they are reached, select whenever there is an option those paths which will give him the best introduction to this wonderful land. . . . At Salina he halts his train for a day while he rides a few miles up the creek to see the unconformity [figure 7-2] between the Tertiary above [i.e., about 55 million years old] and the Jura-Trias and Cretaceous below [minimum age about 75 million years].

Obviously, Gilbert felt that no self-respecting geologist would want to miss this opportunity to have some fun and to see a 55 million-year-old erosion surface. The time-bag represented by the horizontal line on figure 7-2, separating the tilted beds from the

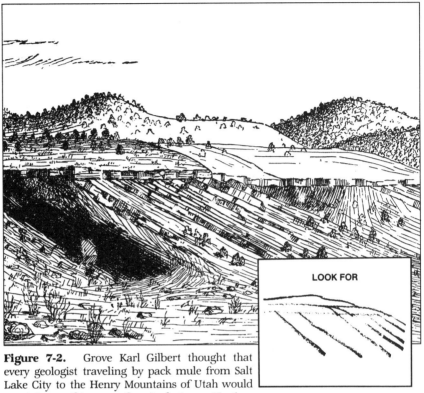

LOOK FOR

Figure 7-2. Grove Karl Gilbert thought that every geologist traveling by pack mule from Salt Lake City to the Henry Mountains of Utah would want to see this unconformity between Tertiary and Mesozoic sediments.

flat-lying ones, contains about 20 million years of tilting, uplift, and erosion.

There are many different kinds of unconformities. One of my favorites is beautifully exposed in a three-hundred-foot-high, wave-cut cliff on the shore of the North Channel of the Irish Sea in County Antrim, Northern Ireland. The wavy white line drawn along the face of the cliff in figure 7-3 marks the position of the unconformity separating a platform of sparkling white, upper-Cretaceous, marine chalk beds from dark-colored basaltic lava flows of Tertiary age. Think of the white line as the edge of a thin time-bag containing

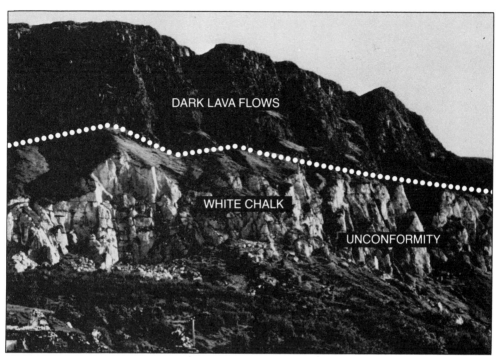

Figure 7-3. This sea cliff on the coast of County Antrim in Northern Ireland exposes a magnificent unconformity between upper Cretaceous chalk and early Tertiary basalt lava flows. Notice how the lavas follow the irregularities of the older topography buried beneath them.

about 20 million years of uplift and erosion. During that interval the chalks were raised above sea level and their once-smooth upper surface was carved into hills and valleys by streams that no longer exist. These were the first of the famous green glens of Antrim. Sunlight was eventually blocked and every living thing was either forced to move or die as basaltic lavas from the earth's interior oozed out of hidden vents and spread across the land.

That story is told quite clearly in a full range of primary structures. A geologist who digs away the heather cover at the top of the flat, chalk slopes will find thin brick-red sheets of baked soil and an occasional charred plant fossil. Similar things are quite common in lava country.

Unconformities and the primary structures that define them were first discovered by our famous friend Dr. James Hutton in 1787 on the island of Arran, southwest of Glasgow. This particular contest of Hutton versus nature ended in a draw. He knew what he had seen (figure 7-4), but he didn't know how to interpret it!

"These two different kinds of stratified bodies rise to meet each other; they are somewhat confused at the immediate junction, but some of the sandstone or calcareous strata overlap the ends of the alpine schistus."

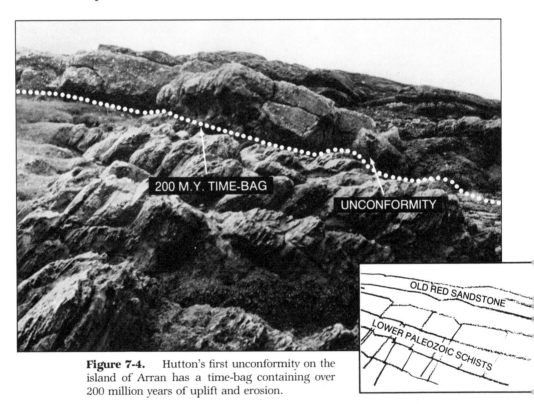

Figure 7-4. Hutton's first unconformity on the island of Arran has a time-bag containing over 200 million years of uplift and erosion.

I know exactly how he must have felt. The first time I visited Arran to see this outcrop for myself I walked blindly past it. The boisterous surf of Kilbrannan Sound was on my left, and a brush-covered sea cliff was on my right. For some reason, I had imagined that the unconformity would be exposed in the cliff. I concentrated my attention on that as I struggled to pick a path through puddles and sopping wet heather on a flat, uplifted terrace. I amused myself by singing the Gaelic name of a local landmark—*Rubha Creagan Dubha* (literally, "point rocky disastrous")—over and over, searching for interesting changes of sound and tempo. After a mile or so I turned back and walked the beach. Success came by surprise when I saw the view of the unconformity shown in figure 7-4.

All three parts of every unconformity are obvious from this perspective. The older rocks, Dalradian Schists of probable lower Paleozoic age, dip to the left. Above them are 360 million-year-old beds of the Old Red Sandstone of Devonian age dipping to the right. The time-bag in between contains at least 200 million years of uplift and erosion. Part of Hutton's problem was that at this stage of his own geologic development, he was unable to appreciate the times that were involved. How soon that was to change.

Hutton knew that there was something interesting about the way certain sets of sedimentary beds rested on older rocks. Therefore, he was determined to explore the question by using the same successful strategy employed at Glen Tilt to discover the relationship between sediments and granites. He would follow a continuous exposure of sediments until he found the contact between them and the underlying crystalline rocks. This time (1788) Hutton chose to follow the sea cliffs along the coastline of the North Sea, a few miles north of the eastern border between England and Scotland. John Playfair (1748–1819) told the story in a biographical sketch of Hutton that was read before the Royal Society of Edinburgh in 1805:

On this occasion, Sir James Hall [1761–1832] and I had the pleasure to accompany him. We sailed in a boat from Dunglass [Hall's estate], on a day when the fineness of the weather permitted us to keep close to the foot of the rocks which line the shore in that quarter, directing our course southwards in search of the termination of the secondary strata [the Old Red Sandstone in this area]. We made for a high rocky point or headland, the Siccar, near which from our observations on shore, we knew that the object we were in search of was likely to be discovered.

Nearly two hundred years later my wife and I decided to join Dr. Hutton and his friends for the landing at Siccar Point. We wanted to climb out on the rocks with them and watch the great men in action. So, we went to Scotland and chartered a fishing boat, the Silver Star, in the picturesque harbor of St. Abbs, about 8 miles south of Siccar Point. The sea and weather were just as John Playfair described them. Scarcely a wave broke on shore, yet the 400-foot-high sea cliffs told us that the stormy reputation of the North Sea was well deserved. I asked our skipper, Captain Mills, about it and he passed my question off with a seaman's flair for understatement: "Aye, but it's a braw day, the new."

As we chugged northward along the coast toward Siccar Point, we marveled at the display of spectacular folds in the 425 million-year-old early Silurian schists. Sir Charles Lyell had used a drawing of these same structures in the shortened version of his textbook, *Elements of Geology* (1838). An appreciation of their appearance may be obtained from figure 7-5 by extending this view for several miles to the left and right in a pattern almost as regular as folds in a kicked rug. It was hard to imagine that we were soon to see these structures covered by flat-lying beds of Old Red Sandstone.

We were in a wonderland. Even the sea birds seemed to know it. Thousands of them fluttered around in constant motion. Puffins and black-frocked guillimots were particularly active. They seemed completely unaware that there is any difference between air and water. Apparently they dive and fly from one to the other without giving a thought to the change. After witnessing that exhibition, our own flight from the 20th century into the 18th to join Hutton appeared to present no difficulties at all.

Figure 7-5. The Silurian schists beneath Hutton's unconformity at Siccar Point are folded like this.

I followed the landmarks on the map as Captain Mills called off their exotic names, Mawcarr Stells, Fast Castle Head and Meikle Poo Craig. Our excitement grew as we rounded the last point and saw the gray colors of the Silurian schists give way to the rusty red-brown of the overlying Old Red Sandstone. We had arrived!

Hutton's unconformity tilts gently northward in such a way that it appears at the top of the cliffs before it can be seen at water level a mile or so farther on. At one place, stone huts of two distinct colors were snuggled against the base of the cliff. Below the unconformity was a gray hut made of Silurian schist. Only three hundred yards away, above the unconformity, was a red hut made of Devonian sandstone. Siccar Point lived up to its billing as a Mecca for geologists. James Hutton had found a classic exposure. Gently sloping beds of the Old Red rest impassively on the upturned and eroded edges of the Silurian schists. Our charter captain brought us in for a landing at the only spot a boat could approach the shore to discharge passengers (see figure 7-6).

Scrambling ashore at Siccar Point is an emotional experience. We felt carried away by its history as we hunted for the same toe- and handholds that Hutton and his friends must have used nearly two centuries ago. We stopped to look around on the rough, wave-cut platform perched about 8 feet above sea level. It is a large surface about 125 feet long and 50 feet wide with an irregular slope down to the sea. A few pedestals, capped by almost flat-lying beds of the Old Red Sandstone, stood on supports of nearly vertical Silurian schistose metasediments. Figure 7-7 illustrates one of these pedestals with the unconformity defined by a dotted white line. Think of this line as the edge of a time-bag containing about 60 million years of folding uplift and erosion.

I traced the position of the buried landscape with my finger as I walked around one of them. The line was easy to follow. All I had to do was keep my finger just above the metasediments and just below the thin conglomerate layer that represents the lowest part of the Old Red Sandstone. The conglomerate layer is only a few inches thick at this point and is overlaid by a foot or so of thin-bedded, red sandstone. Basal conglomerates similar to this one are characteristic parts of most unconformities.

Basal conglomerates are high-energy rocks produced at the point of change from an erosional to a depositional environment. They are often formed by wave action as a slowly advancing sea moves inland, cutting up everything in its way and depositing the

Figure 7-6. We scrambled ashore in this natural harbor at Siccar Point by using the same toeholds that James Hutton must have used in 1788.

fragments in a layer just above the broken edges of the older rocks. Rapidly moving rivers are also capable of introducing coarse gravels and boulders to form a basal conglomerate at the edge of a new depositional basin. At Siccar Point the basal conglomerate is beautifully exposed as a pavement covering an area almost forty feet square. Most of the pebbles are quite angular and represent local Silurian metasediments eroded in Devonian time from local hills and deposited at the beginning of the basin-forming episode.

With all of these things in mind, my wife and I were ready at last to join Hutton and his friends to see how they interpreted the

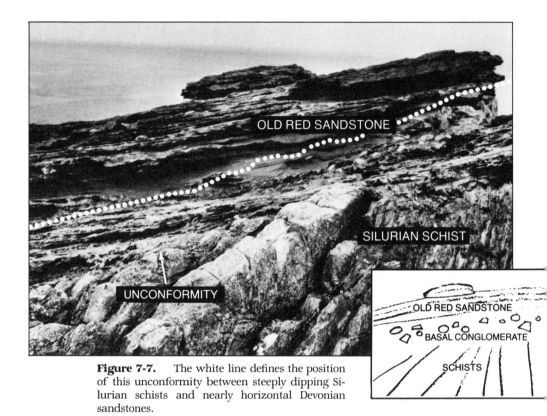

Figure 7-7. The white line defines the position of this unconformity between steeply dipping Silurian schists and nearly horizontal Devonian sandstones.

same set of facts. We sat out of the way, knowing that Hutton would need room to maneuver, while I read aloud from John Playfair's account of that exciting day:

> On landing at this point, we found that we actually trode on the primeval rock, which forms alternately the base and summit of the present land. It is here a micaceous schistus, in beds nearly vertical, highly indurated, and stretching out from the S.E. to the N.W. The surface of the rock runs with a moderate ascent from the level of low-water, at which we landed, nearly to that of high-water, where the schistus has a thin covering of red horizon-

tal sandstone laid over it; and this sandstone, at a distance of a few yards farther back, rises into a very high perpendicular cliff. Here, therefore the immediate contact of the two rocks is not only visible, but is curiously dissected and laid open by the action of the waves. The rugged tops of the schistus are seen penetrating into the horizontal beds of sandstone and the lowest of these last form a breccia [technically, an angular conglomerate] containing fragments of schistus, some round and others angular, united by an arenaceous [sandy] cement.

Dr. Hutton was highly pleased with appearances that set in so clear a light the different formations of the parts which compose the exterior crust of the earth, and where all the circumstances were combined that could render the observations satisfactory and precise. On us who saw these phenomena for the first time, the impression made will not easily be forgotten. The palpable evidence presented to us, of one of the most extraordinary and important facts in the natural history of the earth, gave a reality and substance to those theoretical speculations, which, however probable, had never till now been directly authenticated by the testimony of the senses. We often said to ourselves, what clearer evidence could we have had of the different formation of these rocks, and of the long interval which separated their formation, had we actually seen them emerging from the bottom of the deep? We felt ourselves necessarily carried back to the time when the schistus on which we stood was yet at the bottom of the sea, and when the sandstone before us was only beginning to be deposited, in the shape of sand or mud, from the waters of a superincumbent ocean. An epocha still more remote presented itself, when even the most ancient of these rocks, instead of standing upright in vertical beds, lay in horizontal planes at the bottom of the sea, and was not yet disturbed by that immeasurable force which has burst asunder the solid pavement of the globe. Revolutions still more remote appeared in the distance of this extraordinary perspective. The mind seemed to grow giddy by looking so far into the abyss of time; and while we listened with earnestness and admiration to the philosopher who was now unfolding to us the order and series of these wonderful events we became sensible how much farther reason may sometimes go than imagination can venture to follow.

Playfair was right. The key to understanding James Hutton lies in understanding his controlled use of imagination. Don't think of it as childlike imagination for nothing could be farther from the truth. The imagination of a genius of this magnitude is set on a course directed by detailed observations and controlled by rigorous sequential reasoning. Notice how far we have come on this same path together. We can look at rocks, enjoy their stories and discuss them with similarly inclined thinkers who may have been dead for centuries. That's a big step. It means that we too are using controlled imagination. It also means that we are ready to look into the Grand Canyon of the Colorado River in northern Arizona to see what we can see!

My own first reaction to the Grand Canyon was a tingling shudder of disbelief. I stared down into this gaping hole and it seemed to look back at me with silent indifference, the way countrymen may greet a stranger who enters their crossroads store. The Grand Canyon just lay there impassively, totally uncaring that one more worshiping geologist had peered over the rimrock. The scale of it all was too much for me. That's the moment when tourists are likely to attempt to regain control of their sanity with a wisecrack—"Golly, what a gully!" or something like that. I was too impressed to attempt an escape. I felt it was up to me to meet the canyon's challenge and show it as a reasonably small part of a much larger context.

The Grand Canyon, big as it is—250 miles long, 8 to 12 miles wide, and a mile deep—is very young, not more than 8 to 10 million years old. An older structure would have lost its walls and not be a deep and narrow canyon at all. The Grand Canyon is part of a large machine consisting of five units of simple design: (1) a major river, the Colorado, obtaining its water from the rains and melting snows of (2) the west side of the Rocky Mountains. The river flows westward across a high, arid plateau to a point of discharge in the Gulf of California. The well-being of the canyon depends on (3) the arid climate of the Colorado Plateau (if the climate were humid, many tributary streams would flow into the Colorado River and serve as erosion agents to cut away canyon walls) and on (4) its eight to eleven thousand-foot elevation. No major canyon can be cut below sea level, because there is no way to dispose of the excavated rock. The last element of the design is (5) recent uplift, which prohibits erosion of the walls.

Apparently, the most recent westward movement of North America is becoming blocked by forces below the crust of the Pacific Ocean. At the same time, the subcrustal flow of rock from the center of the Mid-Atlantic Ridge continues to move westward and material is beginning to stack up beneath North America. In quite recent times the entire continent has been elevated enough to expose the Atlantic and Gulf Coastal Plain Province, and to elevate most of the western states. The Colorado Plateau at the common corner of Arizona, Utah, Colorado, and New Mexico has been elevated enough to enable the Colorado River to cut its spectacular canyon.

Topographic contrast between the deepest part of the canyon and the surrounding plateau levels is due to the fact that so much of the erosional energy of the system is concentrated in a path less than one thousand feet wide. Seventy-five thousand cubic feet of water per second loaded with sand and gravel, flowing on an average gradient of about seven feet per mile, have tremendous cutting power. Tributaries originating on the arid plateau with less than ten inches of rain per year cannot keep pace with the great, through-flowing Colorado in the time available to them. It's important to realize that lack of time rather than the abundance of it has been a major factor in creating the grandest canyon on earth. We would have to go to the planet Mars to find its equal!

Now that we've learned to think big, we're ready to take our first look through the vast window provided by the Colorado River and see the hundreds of miles of unconformities exposed on the canyon's walls. A magnificent example, often called "the Great Unconformity," is easily identified in figure 7-8 near the bottom of the canyon at the level marked by the dotted white line. Follow that line around the side of the inner gorge, keeping place between the dark-colored Precambrian rocks of the 1.5 to 1.8 billion-year-old Vishnu Schist below and the light-colored 570 million-year-old Tapeats Sandstone of Cambrian age above. A time-bag spread out along this part of the Great Unconformity contains almost one-fourth of the history of the earth. About one billion years of folding, uplift, and profound erosion are contained in this one episode. Profound is an impressive word, but what does it mean when linked with the word *erosion?* The story is the same as that of the erosion of the Carolina Piedmont.

The Vishnu Schist is a high-temperature and pressure rock type that must have originated some twelve to eighteen miles below what

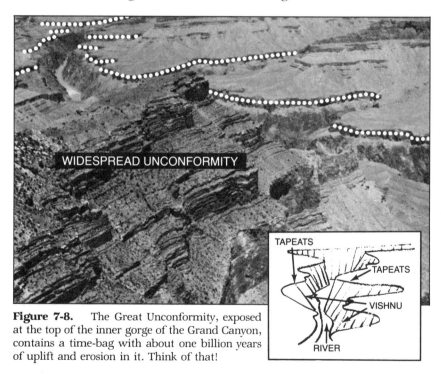

WIDESPREAD UNCONFORMITY

TAPEATS

TAPEATS

VISHNU

RIVER

Figure 7-8.   The Great Unconformity, exposed at the top of the inner gorge of the Grand Canyon, contains a time-bag with about one billion years of uplift and erosion in it. Think of that!

was then the surface of the earth. Therefore, the erosion necessary to expose the Vishnu at this level was equivalent to cutting between twelve and eighteen additional Grand Canyons stacked one above the other like so many giant *Vs* . . . with the walls gone! Now look back at the line of the Great Unconformity in figure 7-8. Notice how the line winds in and out of each side canyon for many miles down river. Convert this line to an image of a large, nearly flat surface of erosion by simply lifting off all the covering rocks that hide its broad expanse from view. Compare this expanded model with the relatively small view obtained by looking down into the Grand Canyon. If your image was large enough—the flat surface must extend at least one hundred to three hundred miles in all directions—you have envisioned the scale necessary for comprehending the significance of the phrase *profound erosion*. The unit is measured vertically in ten to twenty miles and laterally in a few hundred miles. Obviously, the big show at the Grand Canyon is, not the canyon itself, but what can be seen in the walls.

I'll demonstrate what I mean by moving our observation point about fifteen miles up the river and pointing out the features of a more complicated structure. In this place there are two major unconformities instead of just one. Look at figure 7-9 and identify the parts as I list them from the oldest at the bottom to the youngest at the top. The 1.5–1.8 billion-year-old Vishnu Schist is easily recognized at river level. Above that is a tilted continuation of the Great Unconformity, marked on the illustration by a dotted white line. A wedge-like set of rocks, called the Grand Canyon Series, rests above this part of the Great Unconformity. The Grant Canyon Series is probably about 800 million years old and separated from the Vishnu by a time-bag containing another 600 million years of uplift and erosion. The second unconformity between the Grand Canyon Series and the 575 million-year-old Tapeats Sandstone is nearly horizontal. A time-bag at this level contains an additional 200 million years of uplift and erosion. Look carefully, and you'll see that it's all there, arranged as neatly as an accountant's balance sheet. The only thing that's missing is the signature of an auditor.

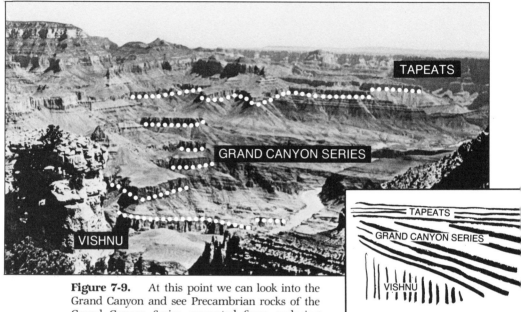

Figure 7-9. At this point we can look into the Grand Canyon and see Precambrian rocks of the Grand Canyon Series separated from enclosing rocks by two unconformities.

We can obtain an audit by moving back downstream a few miles to the place where both unconformities come together and join the single unconformity we saw in figure 7-8. The auditor's part of the canyon wall is shown in figure 7-10. Once again, white dotted lines have been added to identify the positions of the unconformities. Follow the Tapeats Sandstone across the picture from left to right. That marks the 575 million-year time line. Now look just above the pointing finger and identify the portion of the Great Unconformity between the Vishnu and the Grand Canyon Series.

This is the oldest segment of the Great Unconformity that can still be seen. It was originally horizontal and then covered by about two vertical miles of the Grand Canyon Series. These were widespread features extending for vast distances in all directions. Uplift on the left (southwest) and downward tilting on the right (northeast) caused one part to be raised above sea level and lost to erosion

Figure 7-10. Two unconformities come together just to the left of the tip of the pointing finger.

while the other part was preserved. The second unconformity, still to be seen across the entire region, came into being when the area was covered by the Tapeats Sandstone.

Additional facts obtained from many thousands of deep wells drilled while exploring for oil and gas have widened our understanding of these things. The Great Unconformity is known to extend across much of the continent, telling of a time when our land stood high above the sea, much as it does today. Even so, the Grand Canyon remains as the ultimate cathedral of time.

Unconformities made by advancing seas are amazing. As the ocean moves inland across a slowly sinking landscape, wave action cuts away the hills and bevels the surface to form a remarkably smooth plain. At the same time, a set of covering sediments is laid down to blanket the new continental shelf. On lands of low relief near the mouths of rivers, the covering blanket will include deltas, sandbars, and dune deposits as well as the true marine sediments farther off shore. The fall line of the eastern seaboard of the United States is such a feature, marked by numerous waterfalls and rapids as rivers tumble from the Piedmont landform province down to the coastal plain.

Figure 7-11 places these features in the broad context of their formation. The scale is immense. This line, dividing the edge of the continental shelf of Cretaceous and Tertiary ages from the interior

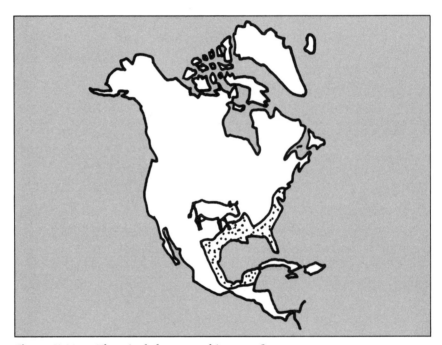

Figure 7-11.   The stippled area on this map of North America represents the Atlantic and Gulf Coastal Plains made of flat-lying Cretaceous and younger rocks that lie unconformably on older rocks. The cow represents many real animals, bovine and human, who step across this sharply defined, four thousand-mile-long boundary without seeing it or understanding how it was formed by erosion and deposition.

uplands, is about four thousand miles long. It stretches in an almost unbroken trace from Long Island, New York, to the Yucatan Peninsula of Mexico by way of Montgomery, Alabama, southern Illinois, south central Oklahoma, and the Big Bend of the Rio Grande. The only breaks in the line are due to burial under a thin veneer of younger river sands. In most places the trace is so sharp that a cow could stand on the older land surface and reach across the unconformity to place one hoof on the emergent continental shelf. Replace the cow with a geologist, map in hand, and the earth begins to reveal its secrets.

That's exactly how the line in figure 7-11 was drawn. Geologists have walked the entire way, mapped the earth and its unconformities, and thereby developed a geological time scale that gives us a more precise context for perceiving where we belong in the scheme of things.

EIGHT

Measuring Time

"Since the time dimension is central
to geology, quantitative measurement
of time is necessary, if earth processes
are to be fully understood."

J. LAWRENCE KULP
"The Geological Time Scale"

William Smith (1769–1839) of Churchill, Oxfordshire, England, taught us how to measure time in a most unusual way. His discoveries led to our ability to identify rocks of equivalent age on all continents. In short, he found a use for fossils.

At age eighteen, Smith was apprenticed to a surveyor and trained as a civil engineer and canal builder. One of his first major jobs was supervising construction of the Somerset Coal Canal near Bath in southwestern England. This was early in the Industrial Revolution, when demand for coal far outstripped the technology to move heavy loads on the rutted wagon roads of the time. Railroads had not yet been invented, and canals were the most economical way to open the countryside for development. The Somerset Coal Canal was designed to join a number of mines to larger canal systems linking the markets of Bristol, Bath, and London. These were busy times for surveyors and practical engineers. Men with training and ability were in great demand. Smith had chosen a good profession for which he was well suited. Nevertheless, beneath his no-

nonsense professionalism was an imaginative and excitable scientist in the midst of formulating an astounding idea.

On the evening of the fifth of January, 1796, Smith returned to his room at the Old Swan, a public house in the village of Dunkerton, and wrote this note:

> Fossils have long been studied as great Curiosities collected with great pains treasured up with great Care and at great Expense and shown and admired with as much pleasure as a Child's rattle or his Hobbyhorse is shown and admired by himself and his playfellows—because it is pretty. And this has been done by Thousands who have never paid the least regard to that wonderful order & regularity with which Nature has disposed of these singular productions and assigned to each Class its peculiar Stratum.

Looking at the same fossils that thousands before him had regarded as "great Curiosities," Smith was stunned by the realization that he was to be the first person to appreciate how to *use* fossils and that, as a result, he was to become a man of destiny. Having found the key to identifying fossiliferous sediments of the same age in all parts of the world, his destiny would lie in mapping the surface rocks of nearly all Great Britain.

By 1814, William Smith had produced a map, entitled "A delineation of the strata of England and Wales with part of Scotland," with a scale of five miles to the inch. His approach was so sound that it has been retained to this day by geologists locating outcrops, designating units, walking contacts, and drawing maps.

Smith went to Somerset as a surveyor, but during the first four years he was there he became a geologist. The transition was made, at least in part, because he had a most remarkable teacher and found himself in a most remarkable laboratory. By strange coincidence, this part of Somerset is the same place described in John Strachey's geological cross-section published in 1719 (see chapter 1).

Smith was in the field for four years before he wrote his note in the Old Swan. He had seen all of these rocks and had compared

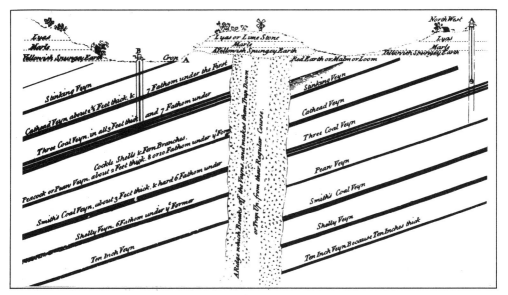

Figure 8-1. John Strachey was William Smith's teacher. He published this cross-section of the Somerset coal beds in 1719. Notice that he even identified the unconformity between the steeply dipping coal measures of Carboniferous age and the overlying Triassic and Jurassic sediments.

them with Strachey's drawings (figure 8-1). He knew the attitudes of the beds and where to find each one. This is probably what led to his realization that fossils belonged in particular beds and not others.

I didn't learn to appreciate the genius of William Smith until I went to Somerset and began tracing the line of his canal around the hillsides. The course of the canal was kept horizontal, stepping up and down the hills in a series of stone-walled locks. It was easy to see how much he could have learned from those long, fresh exposures cut through the lush vegetation and thick soil cover that give Somerset its mellow appearance.

Smith had stumbled onto a great truth but lacked the background to appreciate it. It's one thing to discover that nature has assigned each class of fossils to its peculiar stratum but quite another to know why this is so. The answer involved organic evolution. Progressive changes from species to species through time leave distinctly different fossil records for each age. Every fossil assemblage is equivalent to a date that's stamped indelibly on the containing rocks. It says, "Geologists, take note: these layers were made during the evolutionary lifespans of each species you find here. None of these species will ever be found in rocks older than their moments of evolution or in rocks younger than their times of extinction." Verification of this principle of faunal and floral succession was established beyond all doubt by the work of Charles Darwin and his followers after publication of *On the Origin of Species*. (The link established between a *dicynodont* tooth found in Antarctica with similar fossil teeth found on other continents is a good illustration of how the principle of faunal succession is used.)

Obviously, this concept of the uniqueness of fossil forms does not take in all of the variables that are found in nature. Many factors, in addition to those of evolutionary change, control the types of animal and plant fossils found trapped in rocks.

Depositional environments are particularly important. Every freshwater river entering a brackish tidal estuary and then discharging into the open, saline ocean has its own distinctive biota in each environment. Nevertheless, William Smith's breakthrough was valid. We have simply improved on it over the last two centuries with additional kinds of understanding.

By far the most important by-products of Smith's discovery were, first, the distinction and then, the gradual development of three new kinds of geologic units: *rock units, time units,* and *time-rock units*. An intercontinental geologic time scale emerged from field research as soon as these three units were put to work. These units are illustrated in figure 8-2, showing part of the Little Bahama Bank where a new geologic formation is photographed in the act of being deposited on top of an older one.

We are looking down from an airplane through crystal-clear seawater at a drifting mass of sand made of broken shell fragments and chemically precipitated pellets of calcium carbonate. All of the bright, wavy lines are crests of subaqueous sand dunes. Ocean currents, functioning much like winds, are drifting the loose particles toward the lower right-hand corner of the picture. The dark area is an older and deeper platform of the same material partially hidden under approximately fifty feet of water. For comparison, think of the Tapeats Sandstone as a rock unit that looked something like this nearly 600 million years ago when it was being spread across the surface of the Great Unconformity in what is now the area of the

NEW WHITE SAND
LAYER ABOVE

DRIFT

DARK LAYERS BELOW

Figure 8-2. We are looking down through sparklingly clear seawater over the Little Bahama Bank where a new geological formation is being drifted across an older one.

Grand Canyon. With these new images of rock units spread before us, we're ready to take up the parts one by one.

The basic rock unit is called a *formation,* defined as any mass of rock with readily recognizable characteristics and boundaries. These must have been properly described and given a name associated with a type locality. For the purposes of this book we may consider the drifting carbonate sands of figure 8-2 to be the *Little Bahama Bank Formation* with a type locality twenty miles northeast of Freeport, Grand Bahama Island.

The basic time unit for this formation is called the Holocene—from Greek *holo* ("entire") and *kainos* ("recent"). The age can be established three ways. The enclosed seashells belong to living species. By the law of faunal succession this is a contemporary, or recent, rock. We can also see that this is a recent rock because it is still being formed, as shown by the way it's drifting across the underlying platform. Furthermore, we know that the Holocene began between eleven thousand and fifteen thousand years ago when the great continental ice sheets began to melt away. Meltwater, returning to the ocean basins and spreading over the old platform, made living room for the accumulation of this formation.

Figure 8-2 illustrates a rock unit and a time unit; all that remains is to demonstrate the existence of a time-rock unit. For our purposes, a time-rock unit is defined by a combination of a particular set of rocks and the time-bag that encloses them. Professional geologists express this thought differently. They consider a time-rock unit to be a set of rocks and the time span of their origin. Even at the turn of the nineteenth century, followers of William Smith recognized that the ability to identify the same sets of rocks in different places gave them the power to identify the same time spans as well.

All that remained was to survey the earth, identify time-rock (i.e., time-bag) units, give them identifying names, place them in chronological sequence by the law of superposition and the principle of faunal succession, and construct a worldwide geological time scale. The whole job, started by Smith in 1796, was finished by 1879 with only the need to refine the boundaries of small-time units left.

(The more precisely we can measure things, the more precisely we can think about them.)

The effort to assemble a time scale (figure 8-3), began in a probing, stumbling way. Coal-bearing, continental and associated marine sediments of Belgium and Great Britain were enclosed in a time-bag called *Carboniferous* time. That success bred an additional time-bag study of the chalky beds of Great Britain, Belgium, France, and the Netherlands; this time unit is called the *Cretaceous*. Two more time-bags, the *Tertiary* and *Quaternary*, were soon added to the list. (The basis for judging whether or not rocks belong in these two categories is very interesting. Two criteria are used. The first is the position in space near the top of the rock column. The second is the percentage of fossil forms with living descendants of the same species.) An additional time-bag distinction was recognized in central Germany and called *Triassic* time on the basis of three readily distinguishable units. All of these time-bag surveys were made and published between 1808 and 1834.

By this time, the success of geologists had been so exciting that they realized that a genuine revolution of thought was in progress. The time-rock search became even more intense. *Cambrian* and *Silurian* time-bags were recognized in England and Wales by two researchers, Charles Darwin's old mentor Adam Sedgwick and Roger Impey Murchison, who approached one another from opposite directions. A bitter controversy sprung up because there was no clear-cut place in the fossil record to serve as the upper limit of Cambrian time and the lower limit of Silurian time. Thomas Mann put this problem very nicely in his novel *The Magic Mountain:*

Time has no divisions to mark its passage.

Mann's point is well taken. Every tick-tock distinction that marks off the passage of a second is external to the unceasing flow of time. Without the external marker we have no record of the flow. People who are very time-oriented carry watches while those of us who have less concern for tight schedules feel very little need for them. Professor Adam Sedgwick of Cambridge University and Roger Impey Murchison, a man of private means, were taken aback by the way nature had treated them. Their argument was not settled until 1879 when a third geologist, Charles Lapworth, suggested a compromise between the two time-bags by recognizing a third and intermediate time-bag to be called *Ordovician* time.

The *Devonian* time-bag was defined in southwestern England. It's one of the time units with which we are the most familiar because we have seen rocks of Devonian age above the unconformity at Siccar Point on the eastern coast of Scotland. You'll recall that Dr. James Hutton and his friends John Playfair and Sir James Hall took us there in a boat. That exposure is particularly important because it emphasizes why time units were first thought to be bounded by unconformities. Other exposures in different parts of the world forced a more realistic view of geologic history. A case in point may be taken from the rocks of the Jura Mountains between France and Switzerland.

These were placed in the time-bag that we now call the *Jurassic*. However, that doesn't mean that the Jura Mountains are the only place where rocks of this age are found. Far from it. Both John Strachey and William Smith recognized marine fossils of this age above the unconformity (see figure 8-1) in Somerset. Furthermore, the red rocks standing on end at the edge of the Rocky Mountains and serving as a backdrop to the campus of the University of Colorado at Boulder are also of Jurassic age. Zion National Park, in southwestern Utah, offers another beautiful view of Jurassic rocks. Here, the Navajo Sandstone, with its intricate, wind-blown, sand dune patterns, indicates that the early Jurassic of this part of North America endured a desert climate. Marine invasions and marine fossils came a little later.

The first recognition of the *Permian* time-bag was made in Russia in the Perm area just east of the Urals. That, too, is a time unit with representative rocks in many parts of the world, including the great oil-producing Permian age basin in West Texas.

Mississippian and *Pennsylvanian* time-bags are quite distinctive in the United States where boundaries are defined on fossil evidence enclosed in the rocks of the Mississippi Valley and the states of Pennsylvania and West Virginia. These terms are not applicable on a worldwide basis because the fossil record is not the same on other continents. Carboniferous is a much better term. Nevertheless, the confusion is important because it emphasizes something we need to know about the arbitrary character of the geological time scale.

The original fieldwork was done primarily in Europe on the basis of natural divisions in the fossil forms found there. If the history of science had been different and some other continent had been the center of geological thought, a workable time scale would have been produced and would have reflected those local patterns. Fossils would have been found to support the choice of each time marker, and the overall organization of the time scale would have been similar to the one we now use.

The geological time scale (figure 8-3) is organized in a very simple way. The largest time units are called eras. There are four of them. The *Precambrian* era is the oldest. It contains all time from the creation of the earth to the first appearance of the trilobites. The *Paleozoic* era—from the Greek *palaios* ("old") and *zoo* ("life")— opens with the first appearance of the trilobites and ends with the first appearance of the dinosaurs. The *Mesozoic* era—from the Greek *mesos* ("middle")—is defined by the total life span of all the dinosaurs; it begins with the appearance of the dinosaurs and ends with their extinction. Our present era, the *Cenozoic*—from the Greek *kainos* ("recent")—contains all time since the extinction of the dinosaurs. This is an ever-extending era that becomes longer as each arriving second passes into history. It's interesting to note how much the romantic influences of the mid-nineteenth century pervade this classification scheme.

Dinosaurs—from the Greek *deinos* ("terrible") and *sauros* ("lizard")—are relatively rare in the fossil record; yet because they are such dramatic animals, they were assigned the critical role in fixing boundaries for three of the eras. The other critical role of fixing the boundary between the Precambrian era and the Paleozoic era was also given to a dramatic fossil form, the trilobites (see figure 1-1).

ERAS	PERIODS	EPOCHS
CENOZOIC Mammals abundant	**Quaternary:** in France Desnoyers, 1822	**Holocene** (entire recent): in England Lyell, 1833 **Pleistocene** (most recent): In England Forbes, 1846
	Tertiary: in France and Italy Lyell and Deshayes, 1833	**Pliocene** (more recent): in France Lyell and Deshayes, 1833 **Miocene** (less recent): in France Lyell and Deshayes, 1833 **Oligocene** (new recent): in Belgium and Germany von Beyrich, 1854 **Eocene** (dawn of recent): in France Lyell and Deshayes, 1833 **Paleocene** (old recent): in France Shrimper, 1874
MESOZOIC Dinosaurs	**Cretaceous:** in Belgium, France, and Netherlands d'Halloy, 1822 **Jurassic:** in Switzerland von Buch and von Humbolt, 1839 **Triassic:** in Germany von Alberti, 1834 **Permian:** in Russia Murchison, de Verneuil, and von Keyserling, 1854	
PALEOZOIC Trilobites to amphibians	**Carboniferous:** in England Conybeare and Phillips, 1822 **Devonian:** in Wales and England Sedgwick and Murchison, 1835 **Silurian:** in Wales and England Sedgwick and Murchison, 1835 **Ordovician:** in Wales and southern Scotland Lapworth, 1879 **Cambrian:** in Wales and England Sedgwick and Murchison, 1835	**Pennsylvanian:** in Pennsylvania Rogers, 1858 **Mississippian:** in Mississippi Valley Owen, 1839
PRECAMBRIAN Life rare Record sparce No trilobites	No truly valid worldwide divisions; principally due to lack of fossils	

Figure 8-3. The Relative Geological Time Scale with a list of the men who produced it, the localities where the units were defined, and the dates of first publication.

Each of the periods listed in figure 8-3 is a time-bag enclosing a set of rocks called a *system*. Smaller time-bags, called *epochs*, are included on the chart for the Tertiary and Quaternary Periods because the words have been used throughout this book. No mention is made of additional epochs that have been defined for Paleozoic and Mesozoic time. That much detail would be too burdensome for our purposes.

Think of the assembled geological time scale as a tremendous triumph that was tested before it was finished. The story is thrilling. Charles Darwin's *On the Origin of Species,* published in 1859, explained the relationships between all the various fossil forms and their times of evolution. Darwin's theory established a predicted sequence of events. The emerging time scale and the ordering of the contained fossil record established that Darwin's sequence actually occurred. These two independent approaches proved to be totally compatible. Geologists were very happy at first, then a strange problem arose in 1862.

William Thomson (1824–1907), the famous Lord Kelvin, invented a way to measure geologic time quantitatively in years. Kelvin was a real heavyweight whose opinions could not be dismissed simply because they were inconvenient. He was something of a prodigy with a career that bloomed very early in the field of mathematical physics. His father, James Thomson, professor of mathematics at the University of Glasgow, tutored his son from the age of six until he entered the university at eleven. In 1846, at twenty-two, Kelvin became Professor of Natural Philosophy at Glasgow, occupying the chair he was to hold for the next fifty-three years. His inaugural dissertation, *De Moto Caloris per Terrae Corpus* (*The Flow of Heat Through the Body of the Earth*), was a declaration of war against the geologists' concept of nearly infinite time available for organic evolution.

Kelvin had made a study of the physical history of a white-hot liquid planet as it might be expected to cool down to form the kind of earth we know today. His purpose was to calculate the absolute age of the earth in years based on reasonable assumptions of temperatures, cooling rates, and the residual heat left in the planet today. Kelvin knew from measurements taken in deep mines that there is a three degree C temperature increase per one-hundred meters of depth within the earth's crust. This means that the earth is still leaking heat at the rate of about 2 microcalories per square centimeter per second. From these facts he was able to determine that the age of the earth lay between the limiting figures of 20 and 40 million years. The mathematics seemed to be definitive.

Here was a confrontation of heroic proportions! Kelvin was on one side with unassailable arithmetic, geologists were on the other with their record of an untold number of events that simply could not be crowded into a brief 40 million years. Darwin's position (in *On the Origin of Species*) was stated as clearly as that of Lord Kelvin!

> In all probability a far longer period than 300 million years has elapsed since the later part of the secondary period [i.e., Paleozoic]. . . .
>
> He who can read Sir Charles Lyell's grand work on the *Principles of Geology,* which the future historian will recognize as having produced a revolution in natural science, yet does not admit how 'incomprehensively vast have been the vast periods of time, *may at once close this volume.*

The scientific outlooks of physics and geology were on a collision course. A beautiful illustration of this by two of the participants is given in a conversation between Lord Kelvin and Sir Andrew Ramsay, once director general of the Geological Survey of Great Britain.

> In one of the evening conversaziones [sic] of the British Association during its meeting in Dundee, in 1867, I [Lord Kelvin] had a conversation with the late Sir Andrew Ramsay. . . . We had been hearing a brilliant and suggestive lecture by Professor [Archibald] Gieke on the geological history of the actions by which the exist-

ing scenery of Scotland was produced. I asked Ramsay how long a time he allowed for that history. He answered that he could suggest no limit to it. I said, "You don't suppose things have been going on always as they are now? You don't suppose geological history has run through 1,000,000,000 years?" "Certainly I do." "10,000,000,000 years?" "Yes." "The sun is a finite body. You can tell how many tons it is. Do you think it has been shining on for a million million years?"

[Ramsay said,] "I am as incapable of estimating or understanding the reasons which you physicists have for limiting geological time as you are incapable of understanding the geological reasons for our unlimited estimates." I answered, "You can understand physicists' reasoning perfectly, if you give your mind to it." I ventured to say that physicists were not wholly incapable of appreciating geological difficulties; and so the matter ended, and we had a friendly agreement to temporarily differ.

Temporarily is a relative word. There was no change in these two positions for over thirty years. Finally, after the turn of the twentieth century, physicists began to realize the significance of a series of discoveries that began in 1896 when Henri Becquerel noticed an odd property of uranium. Experimenting with uranium ores, he suddenly realized that they were releasing energy capable of penetrating thick, black paper and exposing photographic plates. He called the property radioactivity—from the Latin *radiare* ("to radiate").

In eight years, through the additional work of Marie and Pierre Curie in France and R. J. Strutt and Ernest Rutherford in England, radioactivity was credited with producing the flow of heat from the earth and with powering the sun in some as yet unknown way. Kelvin's once-devastating arguments that the earth and sun must be very young were swept away. This newly discovered source of energy within the earth invalidated Kelvin's equations based on the rigid assumption that all earth heat was left over from the time of creation.

Physicists at last appreciated that the earth could be very old. But how old? The time had come to be exact. The frontier of historical geology suddenly had shifted from studies of rocks in the field to laboratory studies of the physics and chemistry of radioactive elements. Two important breakthroughs were made between 1902 and 1905. Rutherford and Frederick Soddy discovered that radioactivity was a process of atomic disintegration accompanied by the release of two different kinds of subatomic particles and a form of electromagnetic radiation. The particles proved to be electrons and a combination of neutrons and protons identical in form to helium nuclei. These facts were important because chemical elements are characterized by the numbers of neutrons, protons, and electrons in their makeup. Any loss of these bits of matter from an atom must result in producing a new thing. For example, uranium that has lost electrons and helium nuclei is no longer uranium. It's something else. But what?

In 1905 Bertram B. Boltwood, an analytical chemist at Yale University, found the answer. He had been surprised to discover that every analysis of ores containing as much as two percent uranium also yielded measurable quantities of lead. His cautious conclusion was that there was a mother-daughter relationship between uranium and lead. The next step was dramatic: Rutherford estimated the rate of change from uranium to lead and, by measuring the amounts of each material in a sample, was able to pronounce a date for the original formation of the mineral specimen. But a great deal more had to be learned about the physics of radioactive decay and about the geology of radioactive minerals before the dating method could become a practical tool.

Radioactive disintegration was found to be a random event that affects the entire population of a group of atoms, changing them at a constant rate regardless of the external conditions of temperature and pressure. This means that radioactive disintegration within a molten mass of granite at a depth of twenty-five miles within the earth proceeds just as it does within an outcrop of solid rock buried in glacial ice. A fixed percentage of the population of atoms disintegrates in the same way during each unit of time. The result of this kind of decay is expressed in an interesting property called *half-life*, the amount of time necessary before half of a given population of similar radioactive atoms have decayed to form atoms of another kind of element. A typical half-life decay curve is shown in figure 8-4.

This picture is an important one for it demonstrates the concept of half-life more vividly than words can. Notice the concave shape of the curve. The left edge is high, and the right edge is low, approaching

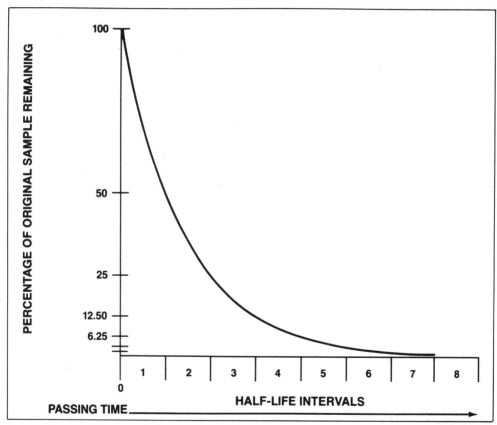

Figure 8-4. This typical decay curve demonstrates the principle of half-life. Half of the radioactive material is altered from the mother element to the daughter element during each half-life interval.

the horizontal line at a continually changing slope that will never permit it to reach the horizontal axis. The shape is the result of losing half the sample during each half-life time interval. Starting with 100 percent of the sample at the upper left, half is gone at the end of the first interval. Half of the remaining 50 percent is lost during the third interval. In the fourth interval, an additional 12.5 percent is lost, and so on to infinity. In a very practical sense, there must always be some

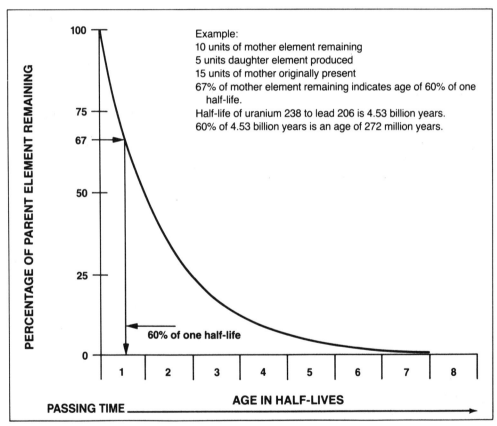

Figure 8-5. This is a half-life decay curve modified to measure the age of radioactive minerals.

of the original material left no matter how long the process has been going on. This is the reason that there is still some uranium left on earth after 4.5 billion years. It also explains why we have so much of the daughter product, the heavy element lead, available to throw at each other during wars and at any unfortunate deer, tigers, and ducks that happen to attract our attention.

The half-life decay curve is modified in figure 8-5 so that it can be used to measure the ages of minerals containing radioactive elements. Picture a single crystal of a mineral containing some uranium bound up in the interior of a great mass of granite. As the uranium decays and becomes lead, both mother and daughter atoms are kept within this crystal cage. Age is the only factor that controls the amount of radiogenic lead to be found. Old specimens contain a larger proportion of lead than do young ones. This transition from uranium to lead takes place on an atom-by-atom basis. Therefore, it is a simple matter to collect the material, divide it into its two mother-daughter components, and calculate the percentage of the original parent material remaining unchanged in the crystal. For example, ten units of uranium and 5 units of lead indicate that fifteen units of uranium were trapped in the specimen at the time of origin. The percentage of the original mother uranium remaining is ten-fifteenths or .67 percent. Knowledge of this value allows us to read the age in half-life units directly from the curve as shown by the arrows drawn on figure 8-5. That value is six-tenths of the 4.53 billion-year half-life of uranium 238, equivalent to an age of 272 million years.

The mechanics of achieving dates of this sort is essentially the same for all the different kinds of radioactive materials. Only the half-life values differ in a stupendous range from 49.9 billion years for the change from rubidium 87 to strontium 87, through 13.89 billion years for the transition of thorium 232 to lead 208, down to 5,570 years for the decay of carbon 14 to nitrogen 14. Geological factors of rock dating are actually more complicated and less certain than the physical ones.

Two critical assumptions are made when the radioactive dating

Eras	Beginning Dates (in millions of years)	Periods	Durations (in millions of years)	Epochs	Beginning and Durations (in millions of years)	
Cenozoic	2	Quaternary	(2)	Holocene	.011–.015 (i.e. 11,000–15,000) (.011–.015)	
				Pleistocene	2	(2)
	70 ± 2	Tertiary	(68)	Pliocene	12	(10 ± 1)
				Miocene	25	(13 ± 2)
				Oligocene	40	(15 ± 2)
				Eocene	60	(20 ± 2)
				Paleocene	70	(10 ± 2)
Mesozoic	135 ± 5	Cretaceous	(65)			
	180 ± 5	Jurassic	(45)			
	225 ± 5	Triassic	(45)			
Paleozoic	275 ± 5	Permian	(50)			
	350 ± 10	Carboniferous	(75)	Pennsylvanian	315	(40)
				Mississippian	350	(35)
	400 ± 10	Devonian	(50)			
	440 ± 10	Silurian	(40)			
	500 ± 15	Ordovician	(60)			
	600 ± 20	Cambrian	(100)			
Precambrian	about 4.5 billion years	(about 4 billion years)				

Figure 8-6. The Calibrated Geological Time Scale. (Courtesy, Geochron Laboratories, Inc., Cambridge, Mass.)

method is applied to rocks. The first is that the ages of the mineral crystals used in the study are the same as the age of the whole rock. The second is that each mineral crystal has been a closed system throughout its entire history. These points need to be amplified.

The best geologic clocks are found in unweathered igneous and metamorphic rocks, glauconitic sandstones, beds of volcanic ash, and black shales. These are the materials that satisfy the first assumption without violating the second. Glauconitic sandstone is an interesting example. Glauconite is a strange mineral formed on the sea floor during the time of deposition of the enclosing sand grains. This mineral is capable of absorbing radioactive elements directly from seawater and holding both mother and daughter atoms without loss for long periods of time. Thin black shales and beds of volcanic ash are also particularly useful geologic clocks. They often contain datable radioactive material deposited in a brief span of time and associated with the same marine fossils on which the Relative Geologic Time Scale was based.

The Calibrated Geologic Time Scale shown in figure 8-6 was assembled bit by bit from many thousands of observations made around the world on rocks of all ages and in a wide variety of geologic settings. Its significance is tremendous. We actually know when the Cambrian began and when the Eocene ended. We know the duration of the Silurian and the approximate time our hominid ancestors evolved from our hominoid ancestors. All that remains is to refine the numbers and narrow the ignorance gaps. After that we can concentrate on the new frontier of discovering the rates of geologic change. This is the frontier that offers an opportunity for creative research. Geologists are excited about that.

In the next chapter we're going to wrap up all we've learned about geologic time, place, process and history and put it into a single context that can be seen no matter where you are on the face of the earth. This is a gigantic wonder story that took fifty years to develop, because most geologists thought that the basic ideas were too astonishing to accept.

PART 3

Seeing
for Yourself

NINE

Dance of
the Continents:
Plate Tectonics

"Isn't it astonishing that all these secrets have been preserved for so many
years just so that we could discover
them?"

ORVILLE WRIGHT
Miracle at Kitty Hawk

The Icelandic pilot of our small chartered aircraft spoke
perfect English. "Look down there at the crack between
those two telephone poles. The last time an earthquake
shook this part of the island, those two poles were moved
apart about fifteen centimeters [six inches]. The linemen had to add
more wire in order to restore service. What do you think of that?"

I whooped a Rebel Yell, the ancient Celtic shout of exultation.
We were right on target over a cracked crest of the Mid-Atlantic
Ridge. This is what my wife and I had come to Iceland in June 1968
to see: some of the realities supporting an important geologic theory
called *plate tectonics.*

The basic idea of this theory is that the crust of the earth is
broken into a number of major plates that move independently and
interact with one another along conflicting borders. Plates that move

197

in opposite directions separate from one another and leave ever-widening cracks, troughs, and even ocean basins in the gaps. Our pilot had just pointed out one of these cracks.

Imagine the thrill of looking down on a set of open gashes in the face of the earth and appreciating that these are part of a continuous fracture system that extends in an unbroken line for forty-four thousand miles. This set of cracks begins at a point on the floor of the Arctic Ocean not far from the North Pole, follows a circuitous path across the floor of every ocean basin, and ends near the west coast of Canada. Most of the cracks are hidden below the sea, but the pattern is exposed beautifully on a north-south line through central Iceland. Great things are stirring within the earth, and the opportunity to see them is heady wine for geologists. Seeing things for ourselves adds zest to life and lets us know that we are sensitive to the ebb and flow of the universe.

Our pilot took us over the vast sand and gravel delta of a spectacular, milky-white river that carries meltwater and sediment to the sea from the island's largest glacial icecap. Parallel sets of linear cracks cut across the top of the newly formed delta and made us realize that these were truly current events. Yet I also felt a sense of history as I remembered the *lystrosaurus* fossil Jim Jensen found at Coalsack Bluff, proving that a crack had appeared, then widened with a motion that separated Africa and Antarctica long ago. Here we were thousands of miles from Coalsack Bluff, staring down at a more recent stage of the same motion.

The sense of history was deepened by appreciating that we were flying within a forty-mile-wide supercrack bounded on either side by high plateaus and steep cliffs. Here was geology at its best. The story of the present and past was laid out to be read. All I had to do was to put the parts together properly.

The little cracks, such as the one between the telephone poles, are simple enough to read. They represent inches of movement as the continents of North America and Europe are separated by subterranean forces. Looking at them is a good deal like looking at fresh rabbit tracks after a snowfall. They're contemporary. The forty-mile-wide crack cut into the lava plateaus east and west of our flight line was something else. Its walls originally had been joined but now were separated by progressive movement over a period of about two million years. I was looking at a stretching rate of approximately 1.2 inches per year. That doesn't sound impressive until you realize that the effects of this motion can be seen by anyone who cares to look at the data and put it in context. My wife and I were going to go back over this same area again and see it at ground level just to be sure we hadn't missed anything. However, that adventure would have to wait one more day, until we had the fun of crossing the Arctic Circle.

We flew on for another forty miles to Grimsey Island astride the 66th parallel. It was exhilarating to head north in the spirit of Pytheas toward mythical Thule. The North Pole lay ahead, across the northeastern corner of Greenland. Suddenly I began to feel the anguish that overcomes all travelers as they reach the limits of their time, endurance, equipment, and money. Every journey stops short of the next hill, an insignificant destination until it's reached. How inviting it then becomes. We approached the Arctic door, tilted a wing in salute, then turned back to the airstrip at Húsavík. Tomorrow would be spent on the ground, driving the coast road to explore details of the supercrack and the minor structures within it.

Looking east from the western edge of the lava plateau with the supercrack below us was worth the trip to Iceland. We had found the plateau to be made of many layers of lava spread one on top of the other like a stack of hotcakes. Their broken edges could be seen in the walls of the cliff bounding the supercrack. About two million years ago the present system of tension cracks began to form, and new lava was spilled out of the earth's interior to fill the void. It was all so easy to see. The volcanic cones and new lava flows were down there as evidence of the creation of the new crust that replaces the old. We marveled at the context. This motion had continued unabated for 180 million years resulting in the separation of North America from Europe and the creation of the Atlantic Ocean basin.

We descended into the trough to see the mere "rabbit-track

Figure 9-1. The depression in the crest of the Mid-Atlantic Ridge can be seen to the right (east) of this set of cracks in the Þingvellir area about twenty-five miles northeast of Reykjavík, Iceland.

crack" our pilot had pointed out from the air. A barbed-wire fence paralleled the line of telephone poles. The earthquake had stretched the wire so tightly that I was able to pluck it like a banjo and hear a fairly clear A above middle C. Small parallel sets of open cracks were commonplace here. Small keystone blocks on the nearby delta surface had dropped a few inches, and part of the roadbed had been dropped as much as eight feet. Each unit was a small-scale replica of the great forty-mile-wide keystone block of the supercrack separating the two plateaus. Our observations had been correct. Iceland was splitting into two parts.

Of course, the local geologists, who work for various national

Figure 9-2.   Cracks such as these near Pingvel-
lir are the result of east-west stretching caused by
subcrustal flow under Iceland.

agencies, had established the context for these movements long ago.
Radioactive dates obtained from various lava flows indicate the west
edge of the island is as much as 16 million years old while the east
edge is about 12 million years old. The rocks in between are progressively younger toward the zone of cracking (figures 9-1 and 9-2),
fissure volcanoes (figures 9-3 and 9-4), lava flows (figure 9-5), and
randomly spaced volcanoes.

   Great things are going on underground and the population of
Iceland is taking advantage of it. Rural electrification, powered by
volcanic heat, is nearly complete for every outlying hamlet and sheep
farm. More than 90 percent of the population has natural hot water
piped directly into their homes for heating and bathing. Virtually all
of the new large buildings are constructed of reinforced concrete to
make them earthquake resistant.

Everyone seems to know about the cracks in the ground and where to go to see them. It's impossible to live this close to the crest of the Mid-Atlantic Ridge and not be influenced by the east-west tug of war spreading the earth's crust. Naturally, visiting geologists come here to see this beautiful illustration of the theory of plate tectonics.

Plate tectonics is a dynamic model of the earth that involves virtually all parts of geology. A thick zone of iron-rich, glassy rock below the crust of the earth behaves as a very viscous, flowing material on which the more brittle crust-rock floats and moves. All of the

Figure 9-3. Fissure volcanoes come in all sizes. These small ones follow a tension crack on the moss-covered lava plain north of the Skaftártunga Valley, Iceland.

great systems of continent building and mountain making are related to interactions of the crustal plates with one another and with currents in the hot, moving rocks below them. Figures 9-6 and 9-7 furnish a global view of what is happening on earth today as the various plates are moved about, jostling one another in the process.

Lines and dots on maps and diagrams are of little value unless they can be translated back into real-world observations from which the data were derived. Consider the pattern of double lines on figure 9-6, extending in a winding path from a point near the North Pole to the west coast of Canada. These lines represent the position of an essentially continuous linear valley that occupies what would otherwise be the crest of the world's longest mountain range. The feature was discovered in 1950 by Marie Tharp, an oceanographer-geologist working at the Lamont–Doherty Geological Observatory of Columbia

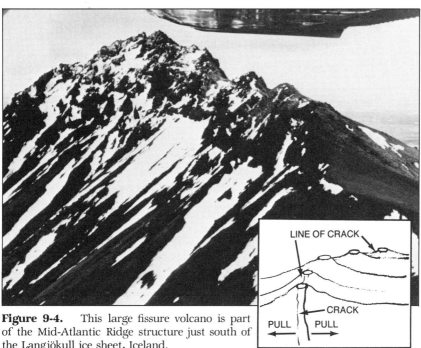

Figure 9-4. This large fissure volcano is part of the Mid-Atlantic Ridge structure just south of the Langjökull ice sheet, Iceland.

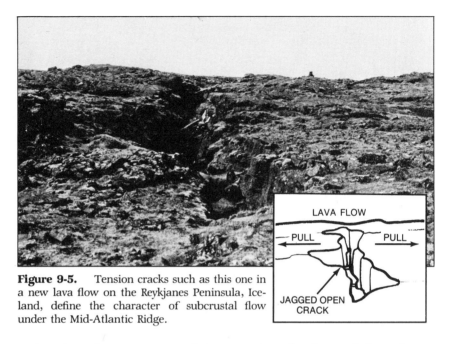

Figure 9-5.   Tension cracks such as this one in a new lava flow on the Reykjanes Peninsula, Iceland, define the character of subcrustal flow under the Mid-Atlantic Ridge.

University. She was engaged in mapping the floor of the Atlantic Ocean and found that the depth charts obtained from six crossings of the Mid-Atlantic Ridge indicated the presence of a valley at the exact center of the mountain range. This gave her the creative idea to search other records for crossings in all the oceans and eventually to find the valley that by explorer's rights should bear her name! The whole thing looks exactly like the central trough of Iceland with the exception that there is a mile or more of ocean water over most of it.

A typical cross section of this ridge system is shown in figure 9-8. Notice how much more genuine the maps and this cross section appear, simply because we have seen something of Iceland and know how the forty-four thousand-mile-long midoceanic ridges must look. Even the earthquake map takes on new meaning because we have an idea how the cracks broke open when subcrustal forces pulled the rocks apart. When we see these things, we see plate tectonics in action. That's what the geological pioneers recognized as they began to assemble the model, bit by bit.

A few highlights of the history of the theory of plate tectonics will be useful to us now. Many of the critical observations that helped bring the idea into focus can be seen by anyone who cares to go to the right places and look for them. Notice how the many facets of geology—time, place, materials, process, and history—come together as we build their contexts. We'll begin with an amusing, early example of plate identification found in an entry (May 23, 1661) in

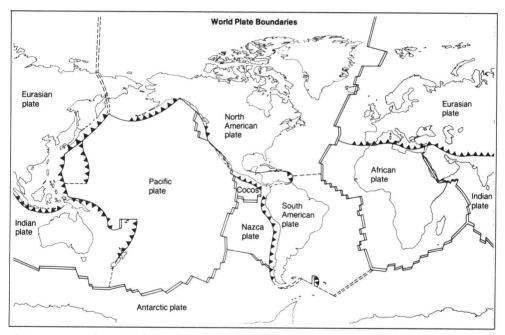

Figure 9-6. These are the major crustal plates of the world. The sets of double lines represent midoceanic ridges where the seafloor cracks and spreads away to plunge down into the body of the earth along the arcs marked by the triangular points. (Reproduced from Sawkins et al., by permission of Macmillan Publishing Co., Inc., New York)

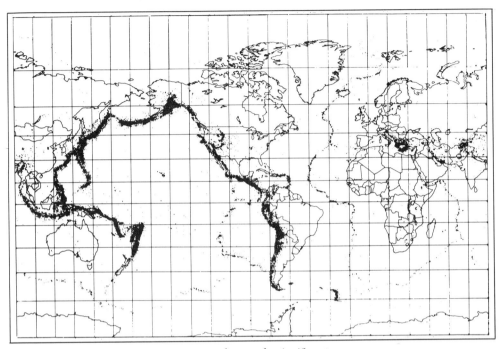

Figure 9-7. Forty-two thousand significant earthquakes that occurred between January 1, 1961, and September 30, 1969, are plotted as black dots on this map. Their distribution shows where rock-breaking forces occur along the mid-oceanic trenches and places where the crustal plates turn down into the body of the earth. (Courtesy, United States Geological Survey, after Barazangi and Dorman)

the famous diary of Samuel Pepys: "To the Rhenish wine house, there came Jonas Moore, the mathematician, to us; and there he did by discourse make us fully believe that England and France were once the same continent, by very good arguments, and spoke of many things, not so much to prove the scripture false as that the time therein is not well understood. . . ."

As surveyor general of ordnance during the Restoration, Moore's responsibilities included map making and the supervision of civil

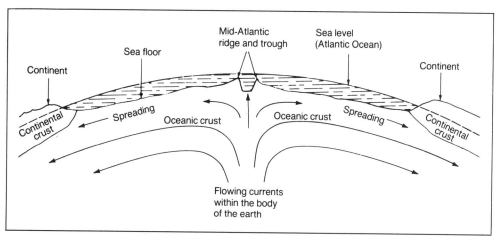

Figure 9-8.   Marie Tharp discovered that the cross section of a midoceanic ridge contained a long central trough where the crest should be. Iceland contains a feature like this that happens to protrude above sea level.

engineering projects. One of his most ambitious efforts was to complete the drainage of the fens, a large, fertile area of glacial peat bogs in East Anglia. Apparently the Dutch engineers, brought in to help with the work, convinced Jonas that Great Britain and Europe were isolated parts of a single structural unit. This is true enough. Great Britain was separated from Europe by the valley of the Rhine River, which was deepened by glacial erosion and flooded after the ice melted away. Perhaps Moore's ability to recognize such relationships becomes more understandable when we recall that he was as much a romantic as he was an engineer.

The fens have been drained periodically since Roman times. It's great fun to fly over them and try to distinguish the many generations of ditches, sea walls, old Dutch windmills, and modern pumping stations spread below. I have done this with Moore's feverishly sincere engineering poem, published in 1685, running through my mind:

I sing Floods muzled, and the Ocean tam'd
Luxurious Rivers govern'd, and reclaim'd
Waters with Banks confin'd, as in a Gaol,
Till kinder Sluices let them go on Bail;
Streams curb'd with Dammes like Bridles, taught to obey,
And run as strait, as if they saw their way.

Creative people seem to share an uncommon amount of imaginative romanticism. Alfred Wegener, whom we met in chapter 3 as the mover of continents, is a perfect example. Wegener was capable of looking at a map of the world and, after recognizing the jigsaw fit of the continental margins on either side of the Atlantic Ocean, capable of searching for ways to prove that separation and continental drift had occurred. He was also capable of recognizing the absolute necessity for continental displacement to explain the distribution of the glacial deposits and climates of Permian time. Wegener saw stationary things and made them move. His error lay in imagining that continents drifted through seas of basalt in the manner of treasure galleons plowing across the Spanish Main.

Harry H. Hess, the man who actually solved the problems of moving continents, was a different kind of romantic. This Princeton University professor of geology had been a naval officer in command of convoys that sailed remote parts of the South Pacific Ocean during World War II. A practical romantic, Hess seized the opportunity to use the vessels' sounding devices to map the topography of the ocean floor on a wide front. Two pieces of data stirred Hess's imagination.

Up to that time the ocean basins were thought to be billions of years old and to contain a complete sedimentary record. Hess found that the sea floor was much too rough to be buried in a thick, smooth blanket of ancient sediments. He also found that the sea floor is marked by deeply submerged, flat-topped volcanic cones of unknown, but apparently strange origin. This wasn't much to go on, but it was enough to tell him that the ocean basins are not the inert regions geologists always had thought them to be.

Bruce Heezen, another geologist at the Lamont Observatory, expanded on Marie Tharp's discovery by recognizing that her linear valley must be a tension feature. This led him to the discovery that the mid-oceanic ridges were places where the thin basaltic crust of the ocean floors was being pulled apart. H. W. Menard, a former naval officer working at the Scripps Institute in La Jolla, California, saw something else. His mapping of the topography of the sea floor revealed the presence of large fracture systems that crossed the mid-oceanic ridges at right angles. Some of them showed that the ridges had been offset laterally as much as several hundred miles. This was evidence that the force systems that produced the ridges extended far into the ocean basins. By 1957 great things were brewing.

On March 26, 1957, the students and faculty of the Princeton Department of Geology assembled in Guyot Hall to hear the then-thirty-two-year-old Heezen speak on the significance of the mid-oceanic ridges. Heezen was convinced that the earth was expanding. However, Harry Hess, who was in the audience, saw something else in the data. After Heezen finished, Hess stood up and said, in effect, "You have shaken the foundations of geology!" Hess knew that scientifically conservative geologists could no longer ignore the fact that continents move and that ocean basins are, in a sense, temporary rather than permanent features.

Four to five years later, a new world view of mobile crustal plates floating away from midoceanic ridges and accomplishing all the purposes of Wegener's concept of continental drift emerged to take its place.

The authors of this model, Hess and Robert S. Dietz, arrived at the idea independently. Hess solved the problem by thinking as Heezen had, from movement at the midoceanic ridges toward continental shorelines. Dietz approached the question the other way. He recognized what happens as oceanic crustal plates interact with continental boundaries and then carried his model out to sea.

We will follow Dietz's approach, because the data on land is there for everyone to see and enjoy. Many of the structures are familiar ones that simply need to be put into this new context in order to take on surprising meanings. An example of this type of thinking is

Figure 9-9.   The processes of sea-floor spreading are visibly at work here on the island of Heimaey, eight miles south of Iceland. Imagine what this volcanic cone will look like twenty million years from now, three hundred miles down the slope of the Mid-Atlantic Ridge toward Europe and submerged beneath a mile of seawater.

to be found in the interpretation we'll place on the volcanic island of Heimaey, located within the fracture system of the Mid-Atlantic Ridge about eight miles south of Iceland.

The view of Heimaey shown in figure 9-9 includes an eroded, extinct volcano in the foreground, the half-buried fishing village in the middleground, and the recently active volcano in the background. Dark-colored lavas and ash falls from the devastating eruptions of early 1973 are still to be seen emitting steam near the top of the cone. What we can't see in the quick wink of the camera's shutter is the cumulative effect of the slow progression of seafloor spreading. The fate of this island is identical to the histories of the flat-topped volcanic cones Hess discovered during World War II. In 20 million years Heimaey may be expected to be drowned under about a mile of seawater and to lie about three hundred miles closer to Europe. The people who live on the island don't fret about that. They are much more concerned with immediate threats of additional volcanic eruptions destroying their homes and closing their harbor.

Most creative geologists make their livings doing relatively routine research, using well-established principles. The creative part of their work is generally a by-product that appears as the result of peripheral vision. Something outside the expected course of things focuses attention on a random observation and suddenly places it in a new context. A perfect example of this may be seen in a 1953 publication about lateral movement on the San Andreas Fault in California. The authors, Mason L. Hill and T. W. Dibblee, Jr., were petroleum geologists employed by the Richfield Corporation to discover oil-bearing structures. What they found was much more exciting.

The San Andreas Fault extends from the Gulf of California for about 600 miles to the shore of the Pacific Ocean near Point Arena, 150 miles north of San Francisco. From there the fault continues north across the seafloor for at least another 160 miles to the Mendocino Fracture Zone. We know its path across the state of California because we can see it; figure 9-10 offers convincing evidence of that. Hill and Dibblee found that a number of pairs of identical rocks of identical ages occur on both sides of the fault and yet are separated from one another by varying distances, depending on their ages. The oldest pair is separated the most; the youngest pair is separated the least. The conclusion is obvious: The San Andreas Fault must be a major break between two moving plates of the earth's crust.

Additional fieldwork has confirmed this conclusion. A very solid piece of evidence for long-range lateral slip on the fault was established in 1976 by Vincent Matthews III. He found ten different characteristics of the Pinnacles Volcanic Formation on the west side of the fault in San Benito County to be almost identical to those of the

Neenach Volcanic Formation in Los Angeles County, 196 miles away. These two rock structures are 23.5 million years old. They must represent a single unit that has been broken apart by fault movement. There is now no question that the state of California consists of two great crustal plates that are sliding past one another at a rate of a little more than a half-inch per year. In more common terms, we may think of the block of rock containing San Diego, Los Angeles, and San Francisco to be moving northward toward Alaska and passing the adjacent part of the continent containing Bakersfield, Fresno, and the entire Sierra Nevada Range.

On the grand scale we can see that the North American Plate has been moved westward about one thousand five hundred miles and shoved over a subcrustal current that is driving the Pacific plate toward Alaska. The most obvious results of this movement have been the creation of the San Andreas Fault and the famous San Francisco Earthquake of 1906. (Our old friend Grove Karl Gilbert described that event for us in chapter 1.) A less obvious, but perhaps more sinister result is that San Francisco is threatened by inevitable, future earthquakes that may strike without warning.

Plate tectonic theory was unknown in the middle 1930s when the great suspension bridges were built in the San Francisco area. The automobile wasn't even a significant reality in 1906 when the last earthquake occurred. We can only hope that the next shock won't occur during a rush hour when many people are on the streets, the freeways and the bridges. Five A.M. is a much better time to play earthquake roulette than 5:30 P.M.! It's one thing to see ancient field evidence of plate movements that have pitted rock against rock and quite another thing to face the same situations as contemporary forces are pitted against mankind.

The maps of plate boundaries and earthquake locations, shown in figures 9-6 and 9-7, define the chief danger areas of the world. The familiar news reports that show crumpled buildings and shattered lives seldom place these events in context as part of the worldwide system of plate movements. One of the unfortunate aspects of public education is the lag between the development of scientific understanding and its appreciation as part of our daily lives. There are exceptions however.

In the summer of 1981 my wife and I, while visiting California, spent a day looking at the San Andreas Fault between Palmdale and Hughes Lake, in the area shown in figure 9-10. We had stopped at a realtor's office for advice on the choice of roads. He turned out to have a degree in geology and a very detailed set of maps of the fault made by the United States Geological Survey. Two especially interesting things came out of our conversation with him. We found out how to locate minor faults by identifying lines of cottonwood trees

AERIAL VIEW
OF THE
SAN ANDREAS FAULT RIFT ZONE
FROM
HUGHES LAKE TO LAKE PALMDALE, CALIFORNIA

Figure 9-10. The San Andreas Fault as it may be seen from the air about sixty miles northeast of Los Angeles, California. All points on the left side of the fault have been moved toward Alaska several hundred miles since late Mesozoic time. Look out, San Francisco! Photomosaic of aerial photographs, courtesy U.S. Geological Survey.

that grew best where fault movement had backed up the ground-water. We also learned to respect the professionals who feel an ethical responsibility to warn their clients about the hazard levels that may vary radically over short distances from this spectacular fault. (California is a great place to learn to be an observer. There is so much more to see there than the wonders of Disneyland.)

PUTTING IDEAS TOGETHER

Robert S. Dietz had been living in California for many years before he published his famous paper, entitled "Continents and Ocean Basin Evolution by Spreading of the Sea Floor" (1961). Living in earthquake territory on an obviously moving plate was great prep-aration for developing the imagination to see that virtually all the earth is mobile. The task was to discover how the earth's machinery fits together and how it operates. Dietz's solution was to assemble the parts he knew about along the continental margins and then to extrapolate outward to the newly appreciated midoceanic ridges that had become the intellectual property of Bruce Heezen and Marie Tharp. The key to this study seemed to lie with earthquake evidence, for it marked the places where rocks are being broken by strong and steady forces within the earth.

Hugo Benioff, a seismologist at Cal Tech, had published studies that showed a strange set of shear planes descending beneath conti-nental margins occupied by linear mountain chains onshore and deep oceanic trenches offshore. These began at sea just beyond the ocean trenches and curved back under the mountain chains to depths as great as 450 miles. Dietz and many others had begun to realize at this time that there was only one possible explanation for these features.

The rocky crust under the ocean must curve downward and plunge beneath the edge of the adjacent continent (figure 9-11). Earthquakes occur as the diving oceanic plate rubs against the sta-tionary rocks that form the deep marine trough and linear mountain

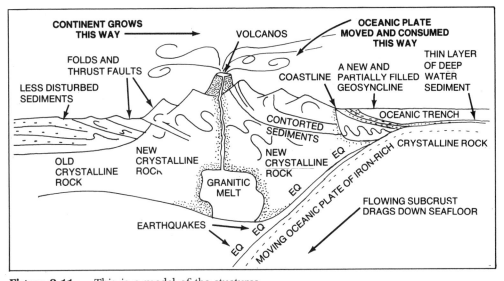

Figure 9-11. This is a model of the stuctures formed at the edge of a continent being underridden by a flowing, oceanic, subcrustal current. These processes are active today around the rim of the Pacific Ocean and the Himalayan belt from the Mediterranean to Java. The triangular points on figure 9-6 show the locations of these structures.

range on shore. There are no earthquakes below 450 miles because the plunging plate is too hot and soft to fracture at that depth.

A beautiful model began to emerge, explaining what had been a perplexing set of scattered observations of the history and structure of mountain ranges.

We may compare this success to that thrilling moment when the parts of a partially completed jigsaw puzzle begin to fall in place. Fingers move rapidly and the pieces take form with a rapid plop, plop, plop! That's exactly the way it was, but on a global scale rather than just a tabletop adventure. Geologists had been accumulating fragments of the picture for millennia and displaying them to one another in discouraging disorder. All of a sudden there was order; everything had its place. Even the blank areas had well-defined shapes that indicated what to look for next. The great geological revolution of the 1960s was detonating! The ideas became infectious and geologists all over the world joined the party. Everyone seemed to want to discover how their own researches could be used to expand the pattern. Plate tectonics became an exciting and very human event.

We'll try to recreate something of the spirit of this revolution by showing how much fun it is to see familiar geologic structures fit together into a single model. Our first example (figure 9-12) shows a forty thousand-foot thick sequence of 1.2 billion-year-old Precambrian sediments called the Belt Series of Idaho, Montana, and northwestern Canada. When we see strata stacked bed on bed in this manner, we should also think of the shape of the original basin in which deposition took place. That thought led James Hall (1811–1898), former State Geologist of New York, to an astounding discovery.

Hall had been mapping the structure of the folded rocks of the northern Appalachians and had measured the total thickness of all the beds involved to be about forty thousand feet. This measurement is many times greater than the thickness of all beds of the same ages in Ohio and in the Mississippi River Valley. His conclusion was that linear mountain ranges are formed from sediments laid down in long, narrow, and deep troughs before being folded up into mountains. We now call these troughs *geosynclines*—from the Greek *geo* ("earth"), *syn* ("together"), and *klinein* ("incline"). They are the fundamental building blocks from which continents are made, unit by unit.

An example of a geosyncline in the process of being filled with sediment and added to a continent is shown in figure 9-11 on the seaward side of the mountain range. The materials are derived from three sources, one on the continent and the other two at sea. Weathering and erosion of the mountains permits sands, gravels, and clays to be carried down to form continental shelf, slope, and abyssal deposits. Precipitation of limestones from seawater and accumulation of shell fragments is particularly common in the warmer

Figure 9-12.   When you see a stack of sediments such as this part of the forty thousand foot thick Belt Series in Montana, think of the deep geosyncline in which they were deposited. This environment is labeled "A New and Partially Filled Geosyncline," in figure 9-11.

parts of the world. The third source of geosynclinal fill is exotic. A thin blanket of sands, gravels, silts, and clays cover the seafloors of the world. As seafloor spreading progresses, some of this material is scraped off the top of the descending plate and is crumpled together as part of the geosyncline. The size and scale of these features are staggering.

Geosynclinal troughs may be several thousand miles long, hundreds of miles wide, and three to ten miles deep. An important part of our developing insight is the ability to envision such large-scale features from rather small samples. Fortunately these systems are so uniform that even small samples are representative of the whole. The fold in figure 9-13 and the thrust fault of figure 9-14 are fine examples.

The great Pacific Ocean crustal plate that is moving down under Alaska has folded and thickened the sediments of the Pacific geosyncline to form mountain ranges. The fold in figure 9-13 is overturned from south to north (left to right) in response to the major movement indicated by the triangular points in figure 9-11. Notice how well the shape of the fold defines both the amount and direction of the crustal shortening that has taken place at this locality. We are looking at a single fold and learning about the direction of the force system and plate movement of the crust of the northern Pacific Ocean. The same story is told even more dramatically at Chief Mountain, Montana (figure 9-14).

The northern Rockies were formed as the westward moving North American plate began to crumple sediments of the Cordilleran geosyncline in late-Cretaceous time. Displacements on many parallel faults with large slices of rock between them served to relieve the pressure as the eastern part of North America moved faster than the western part. One rock slice riding on a three hundred mile-long fault, called the Lewis Overthrust, has brought Precambrian beds of the Belt Series many miles to the east and left them resting on much younger lower-Cretaceous beds. The overthrust cap of Beltian rocks forms the high rampart-like front of the Rocky Mountains in Glacier National Park. Chief Mountain is unusual because the Lewis Overthrust that passes right through it is isolated by recent erosion from the rest of the structure. The point is illustrated beautifully in figure 9-14.

It's fun to look at mountains as museums of the forces that made them. Chief Mountain is quite a display. We can see that the shove that brought the Beltian rocks on top of the Cretaceous shales predated the erosion that separated the mountain from the rest of the range behind it. We can also see the direction of movement

Figure 9-13. This picture of an overturned fold was taken in Alaska by one of my former geology students while flying with a bush pilot. Notice that the rock-bending force came from the left. That is the direction from which the descending Pacific plate is moving down and under the continent and squeezing these geosynclinal rocks up as mountains. Can you fit this situation into the model shown in figure 9-11?

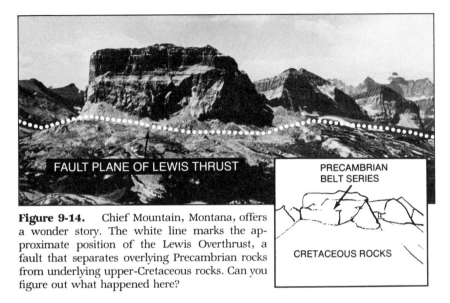

Figure 9-14. Chief Mountain, Montana, offers a wonder story. The white line marks the approximate position of the Lewis Overthrust, a fault that separates overlying Precambrian rocks from underlying upper-Cretaceous rocks. Can you figure out what happened here?

because there is no source of Beltian rocks to the east, beyond the Rocky Mountain front. (I must point out that the Rockies are in midcontinent rather than on the coast. For this reason the picture is a little different from that shown in figure 9-11, but the basic principle of seeing the force system from the direction of offset still applies). Our next insight involves the origins of crystalline rocks.

The high temperatures and pressures found deep within the crumpled geosynclines produce chemical as well as structural changes in the rocks. Shales are converted to slates as the original, iron-stained clays are altered by heat and pressure to feldspars, micas, and dark-colored, iron-bearing silicates. Layer upon layer of gravels, sands, and clays are changed to gneisses and schists, with their easily recognized banded and leaf-like fabrics. The contorted gneiss shown in figure 9-15 is distinguished by light and dark layers of iron-free and iron-rich minerals that reflect original differences in the compositions of the source rocks. Granites form where an environment is hot enough to melt the rocks and destroy nearly all resemblance to the originally bedded sediments from which they were derived.

It's difficult to imagine one of these hard, recrystallized rocks as a porous mixture of gravel, sand, and clay saturated with seawater. Yet that water is the chief reason that heat and pressure are so effective in producing new rocks from older ones. The less stable minerals dissolve in the hot seawater, and then the same chemical elements recombine in new ways to form minerals that are stable at

ORIGINAL
BEDDING
PLANES

Figure 9-15. This gneiss, in the mountains of western North Carolina, was once a set of bedded sands, clays, and gravels before being crumpled deep within the root of the Appalachian Geosyncline. At least twelve to eighteen miles of overlying geosynclinal rock was removed to expose this outcrop. Think of that!

depths of twelve to eighteen miles. Look back at figure 7-9, showing the unconformity between the crystalline Precambrian rocks of the Inner Gorge of the Grand Canyon and the flat-lying Paleozoic sediments above them. The difference between these two sets of rocks is that the crystallines have been subjected to the stresses at the base of a compressed geosyncline while the flat-lying sediments have not.

Sometimes rocks speak quite sharply. Wherever there's sufficient heat, pressure, and water to melt great masses of rocks, they may be expected to intrude the sediments above them. Quite frequently this hot material reaches the surface with explosive results as expanding steam produces violent volcanic eruptions. The plunging Gorda Pacific Oceanic plate, which lies just off the west coast of North America and extends from northern California to the northern tip of Vancouver Island, is responsible for a set of large volcanoes that includes Mount Saint Helens in Washington State. Eastward movement of the plate squeezes the chamber containing partially melted rocks and causes unexpected episodes of volcanic activity for which the local population may be unprepared.

Harry Truman, a longtime resident of Mount Saint Helens, was such a man. His story resembles a classic Greek tragedy of the fourth century B.C., for in him we see the portrayal of a strong man with the so-called "heroic flaw," willful defiance of the forces of nature. Harry chanted his own epitaph to a reporter for *Newsweek:* "No one knows more about this mountain than Harry, and it don't dare blow up on him. . . . If the mountain goes, I'm gonna stay right here and say, 'You old bastard. I stuck it out for 54 years and I can stick it out for another 54. . . .' "

A Greek chorus of scientists came to monitor the events and chant their alarums, but feisty Harry Truman refused to budge. A half century of looking up at the sleeping peak and out across its beautiful slopes, valleys, and lakes had lulled him into forgetting that Saint Helens was classified as an active volcano. After all, by his time scale, eruptions that occurred in 1857 were ancient history.

Earthquakes began on March 20, 1980; an eruption followed shortly after noon on March 27. By April 30, a bulge was noticed on

Figure 9-16. We are looking at Mount Saint Helens in eruption shortly after the big bang of May 18, 1980. That's Mount Hood, Oregon, in the background. (Courtesy, United States Geological Survey, EROS Data Center, Sioux Falls, S.D.)

PACIFIC OCEAN

MOUNT SAINT HELENS

NORTH AMERICA

GORDA PLATE

the north side of the mountain. Things were beginning to happen inside the cone. Sunday, May 18, was the big day. David Johnston, a volcanologist with the United States Geological Survey, was on duty at an exposed observation point, code-named Coldwater 2, when the blast occurred. He had just time enough to radio headquarters a five-word message: "Vancouver! Vancouver! This is it."

Harry Truman put his faith in a bottle of whiskey and an old mine tunnel, in which he intended to hide. Unfortunately, he never had a chance to run. The volcano erupted with a force estimated to have been the equivalent of at least fifteen hundred Hiroshima-style atomic bombs. Prior to the eruption, the mountain had been a beautifully symmetrical peak, 9,677 feet high, which swept up from a broad base in the manner of Japan's Fujiyama. Figure 9-16 shows the partially decapitated cone of Mount Saint Helens and Mount Hood, another Gorda plate volcano, beyond.

One and a half cubic miles of rock and glacial ice were blown away from the side of the mountain, lowering the crest by thirteen hundred feet, and leaving a crater nearly a mile wide. The blast was heard two hundred miles away. Finely divided, glassy rock fragments, called volcanic ash, were blown downwind in a broad swath that covered the continent from Mississippi on the south to the Gaspé Peninsula on the north.

David Johnston's father spoke about it later. "David took risks; a couple of years ago he was trapped on Mount Saint Augustine in Alaska just before it erupted, but they got him out with a few hours to spare. . . . Not many people get to do what they really want to do in this world, but our son did."

I feel very close to David and his situation for two reasons. One of them is that of being kindred spirits I've usually been able to do the things I've wanted to do. The second reason is that of being impressed by volcanoes. The first volcano I ever got to know personally is the 14,408 foot peak of Mount Rainier, about ninety miles north of Mount Saint Helens. World War II was raging at the time and I was stationed in Seattle another ninety miles farther to the north. My unit had been in position for six weeks before the cloud cover broke up and I saw this great cone for the first time. The view is staggering. There was nothing to do but find a geologist and get him to explain it to me.

Dr. George Goodspeed, Chairman of the Geology Department at the University of Washington, listened patiently to my stammers of amazement and then took me to a window of his office to point out the structure of the region. When he reached the looming outline of Mount Rainier he said, "That's only yesterday." My response was an expectant question, " . . . And maybe tomorrow there'll be another eruption?" Goody, as his students called him, nodded and said, "Yes, maybe tomorrow it'll pop off again; who knows? Do you see that peak just below the top of the crater? That's all that's left of an older cone. Rainier blew her top once and has been rebuilt since then."

Big volcanoes produce this kind of feeling of awe. That's exactly the way my wife and I felt when we visited Mount Vesuvius and Pompeii in order to see what is happening as the African Crustal Plate plunges under the European plate in southern Italy. Pliny the Elder had been on the flank of this volcano at the time of the great eruption of 79 A.D. that destroyed Pompeii and Herculaneum (figure 9-17). Pliny's role as an observer, and the David Johnston of his day, was told by his nephew, Pliny the Younger. The old man died with his sandals on because he felt duty-bound to find out what was going on.

> Someone had called his attention to a curious thing:
> . . . [A] cloud which appeared to be of a very unusual size and shape . . . was ascending, the form of which I cannot give . . . a more exact description than by likening it to that of a pine tree, for it shot up to a very great height in the form of a tall trunk, which spread itself out at the tip into a sort of branches; . . . it appeared sometimes bright and sometimes dark and spotted, . . . This phenomenon seemed to a man of such learning and research as my uncle extraordinary and worth further looking into. . . .
>
> He ordered the galleys to put to sea, and went himself on board. . . . Hastening then to the place from which others fled

Figure 9-17. Mount Vesuvius with the ruins of Pompeii in the foreground and a fragment of decapitated Mount Soma to the right of the cone tell us that the African plate is still in motion. Look out Naples!

with utmost terror, he steered his course direct to the point of danger, and with so much calmness and presence of mind as to be able to make and dictate his observations upon the motion and all the phenomena of that dreadful scene.

He was now so close to the mountain that the cinders, which grew thicker and hotter the nearer he approached, fell into the ships, together with pumice stones. . . . Here he stopped to consider whether he should turn back again; to which the pilot was advising him, "Fortune," he said, "favors the brave; steer to where Pomponianus is."

Pomponianus was a friend living in a villa on the slopes of Vesuvius. After a bad night of bombardment, Pliny made his second mistake. He decided the group should go back to the ships. They set out in comical fashion, protected from falling ash by pillows that were tied on top of their heads with napkins. The effort was too much for him and he died of an apparent heart attack after laying himself down on a sail cloth and calling for water. Pliny's story remains as the first scientific account of a volcanic eruption. When seen in context it is also an account of a scientist's stumbling attempt to understand one aspect of plate tectonics.

Perhaps the best way to appreciate the concept of plate tectonics is to try to use it to explain the world we live in. The method is simple. Pick out some feature and see if it seems to fit into the model that we have developed in this chapter. Good agreement and poor agreement must be judged on the basis of what is or is not least astonishing.

The Great Falls of the Potomac River, about twelve miles upstream from Washington, D.C., offers an excellent opportunity to test the method on a feature that may not appear to be related to plate tectonics at all. The view of the falls (figure 9-18 top) from a point below it on the Virginia side of the river is spectacular. This tumbling array of cataracts is astounding to anyone whose idea of the Potomac has been formed at the wide tidewater estuary that can be seen near National Airport. Things are so different up here in this high-energy environment where the river falls dozens of feet and the great wall of rock retreats vigorously upstream. That vigorous retreat is an open invitation to look downstream to see what has happened in the past.

The view shown in figure 9-18 (bottom) is even more astounding. Look at the profile of the river valley. It's wide and flat on top and yet contains a deep, narrow gorge in the center. The contrast tells us that the profile was produced by two different mechanisms operating at different times. The narrow gorge is obviously the product of waterfall retreat as the Great Falls were moved back upstream. Flat valley floors are made by meandering rivers that wander back and forth and widen their valleys. We can see the far wall of this one at the extreme upper left of the picture. One other piece of general information is available to us. Flat valleys are carved that way because the central stream is not cutting down but is flowing on a low slope to the sea.

In this case the sea was less than fifteen miles away. It covered most of the area of Washington, D.C., and, as evidenced by the record of Mid-Tertiary sediments, the entire area of the Atlantic and Gulf Coastal Plains shown in figure 7-11. Figure 9-18 (bottom) now becomes a beautiful picture of land form stability in Mid-Tertiary time followed by a dramatic increase in the radius of the earth under a very large part of North America. Think of what we have done.

Our observation was made at a local spot, the Great Falls of the Potomac River, but it has placed that spot into context with the rest of the continent. Large-scale regional uplift was inevitable after Mid-Tertiary time when the continent was no longer able to move westward as freely as it had before its impact on the deep structures below the Pacific Oceanic plate. Extra rock, moving slowly westward from beneath the Atlantic Oceanic plate, has been moving under the continent since Mid-Tertiary time, increasing the radius of the earth and literally lifting most of North America. The uplift can be seen at thousands of places. The most familiar ones are at the eight thousand to eleven thousand-foot elevations of the Colorado Plateaus followed by the entrenchment of the Grand Canyon. It is also evident at the high plains east of Denver, the Appalachian Plateaus extending from Alabama to Pennsylvania, and, of course, along the Fall Line of the Eastern Seaboard. That's a formidable bit of engineering!

No wonder Dr. James Hutton coined the phrase "the earth machine" as a way of describing the succession of changes observed in nature. We, too, have found the earth to be a bold composition of interactions that seems to remain in a constant state of dynamic equilibrium.

RETREATING WATERFALL

RETREATING

Figure 9-18. *Top,* the Great Falls of the Potomac River, twelve miles upstream from Washington, D.C. *Bottom,* look downstream at the evidence of regional uplift and the story of plate tectonics.

OLD VALLEY FLOOR

NEW CANYON

Epilogue

Picture windows that let nature into a room became practical with the development of thermally insulated double-glazing. Our book is intended to function in the same way, letting nature into your life through the windows of ideas and pictures that reflect realities waiting to be seen outside. The next step is yours. Pass through the looking glass. Go outside and discover the things that are happening around you. Any plowed field, rock, stream, or land form will speak to you in the language you have learned to read. Your apprenticeship is over. You are ready to become a practitioner, capable of teaching others the new things you have been learning, and will continue to learn, for yourself. The technique of learning is simple.

When a journeyman geologist sees a rock, he or she will search out the primary structures that tell how the rock was made. A journeyman studying a flowing stream, wave-washed shoreline, or plowed field will observe what has been moved, for this defines the story of change through time. A journeyman looking at an outcrop or land form will identify the parts and put each one into its proper time-bag. Once these things are done, the new view is complete in four dimensions. Every geologic situation begins to make more sense when we appreciate the wasness of the is. The joy associated with living is quite different for you as a journeyman geologist than it was for the "preapprentice" who picked up this book.

I shall prove that this is true and that you actually have been graduated from your apprenticeship *cum grano salis conditi* (with a grain of spicy salt). Turn back to the frontispiece where I first extended the invitation to "follow me." Nothing has changed in that picture; yet we are able to see it now in a different way because we have become accustomed to looking for context.

Think about Harrington's First Law of Science: nature is scrutable when everything is seen in context. A postapprenticeship reader will try immediately to think of the right questions, beginning with a simple one: Where is this black-suited climber? He's far above timber line on a glacier that gives access to the summit of Mount Rainier, a companion volcano to Mount Saint Helens.

These words have meaning for us now, largely because we can appreciate what they stand for. When we think of the word *glacier*, the nineteenth century research of Louis Agassiz in the Swiss Alps comes to mind. Linking the name Mount Rainier to the phrase "companion volcano to Mount Saint Helens" evokes the story of plate tectonics and the conflict between two moving crustal plates on the rim of the Pacific Ocean. Suddenly we realize that we are seeing and thinking simultaneously. Thinking gives our seeing scope; and peripheral vision. Seeing gives our thinking substance and focus. We are different. That's the point of the apprenticeship.

We cannot see without thinking; we cannot think without seeing. This capacity separates people from one another and from the lower animals. Welcome to the fellowship of geologists *quod tu id meruisti*—"Because You Earned It!"

Notes

Follow Me

William Blake, *The Poems of William Blake* (London: Walter Scott Publishers, 1888), p. 208.

Albert Einstein, *Out of my Later Years* (New York, the Philosophical Library, 1950), page 61.

Chapter 1

John Lear, "The Bones On Coalsack Bluff: A Story of Drifting Continents," *Saturday Review*, 7 Feb. 1970, pp. 47–51. The facts of the Colbert-Jensen story were taken from this article, then verified and amplified by telephone conversations with the participants. Also Edwin H. Colbert, *A Fossil Hunter's Notebook* (New York: E. P. Dutton, 1980). The story of the Coalsack Bluff is included in chapter 15. This account emphasizes the human side of the story.

Rudyard Kipling, *Stalky and Company* (London: Macmillan and Co., 1899), p. 272.

Mary Chandler, Robert Arnold Aubin, *Topographical Poetry in XVIII Century England,* (New York, Modern Language Association of America, 1936) p. 164.

John G. C. M. Fuller, "The Industrial Basis of Stratigraphy, John Strachey, 1671–1743, and William Smith, 1769–1839," *American Association of Petroleum Geologists Bulletin* 53 (1969): 2256–73. A modern geologist who plans to visit southwestern England will find this article to be a door to adventure.

Grove Karl Gilbert, "The Investigation of the San Francisco Earthquake," *Popular Science Monthly* 69 (1906): 97–115. This article offers a magnificent opportunity to observe Gilbert's way of thinking.

Harold W. Scott, *Lectures on Geology by John Walker* (Chicago: The University of Chicago Press, 1966), p. xvii. John Walker was probably a very good teacher. One of his most memorable comments was, "I am teaching a course I have never been taught." He was on the frontier.

Timothy A. Conrad in P. E. Raymon, *Prehistoric Life* (Cambridge: Harvard University Press, 1939), p. 47.

Chapter 2

T. G. Bonney, *Charles Lyell and Modern Geology* (New York: Macmillan and Co., 1895). The dicta that guided Lyell's life are found on p. 213. One of the most interesting facets of this book is a step-by-step description of Lyell's creativity, while developing views on the principle of uniformity and the three-volume text in which they were published. An example of his thinking is given in Charles's letter to his father (1827).

Linus Pauling, personal communication, 1979. This anecdote was given to me in answer to a question on first becoming aware of thinking.

Francois Marie Arouet Voltaire, *Oeuvres Completes* (Paris: Imprimerie et Founderie d'Everat, 1836), vol. 5, *Elements de la Philosophie de Newton*. The original edition was printed in Amsterdam in 1738 only eleven years after Newton's death. Madam Conduit, Newton's niece, gave Voltaire the information based on childhood memories. Physics teachers will find this paper a most stimulating reflection on both Newton and the birth of science.

Arthur Conan Doyle, "A Study in Scarlet," in *the Boy's Sherlock Holmes,* ed. Howard Haycroft (New York: Harper & Row, 1892), pp. 21–22. Holmes uses similar reasoning in a number of other places, notably "The Case of the Greek Interpreter."

Leonardo da Vinci, *Note Books,* ed. Edward Mac Curdy (New York: Empire State Book Co., 1923). Additional information on da Vinci's geological observations is to be found in Kirtley F. Mather and Shirley L. Mason, *A Source Book in Geology* (New York, McGraw-Hill, 1939).

Rudyard Kipling, "The Benefactors," *American Magazine* 74(1912): 259–68. "The Benefactors" is an antiwar poem with a powerful ending.

Joseph Needham, with the research assistance of Wang Ling, *Mathematics and the Sciences of Heaven and Earth* (Cambridge: at the University Press, 1970), (vol. 3 of *Science and Civilization in China*), p. 614. Used by permission of the Cambridge University Press.

James Hutton, *Theory of the Earth* (Darien, Conn.: Hafner Publishing Co. 1970). This is a facsimile edition of three of Hutton's printed works, including *System of the Earth* (1785) and *Observations on Granite* (1794), as well as John Playfair's *Biographical Account of the late Dr. James Hutton.* The introduction by Victor A. Eyles, foreword by George W. White, and Playfair's biography furnish a beautiful portrait of the man and his work. Hutton's trip to Glen Tilt is recorded in the granite paper, pp. 137–38.

John Playfair, *Illustrations of the Huttonian Theory of the Earth* (New York: Dover Publications, 1956), republished as an unabridged copy of the 1802 edition. Playfair establishes the real pattern of Hutton's reasoning by following the mathematician's step-by-step derivation of an idea from the basic facts to the final conclusions.

Charles Lyell, *Principles of Geology*, 3 vols. (London: John Murray, 1830–33). These books revolutionized the science of geology. They served three generations of geologists as a model of fact and reason. These texts became a unifying educational experience that soon bound the geologists' fraternal subculture together. The effect was almost as if Lyell's brain was available to be shared by all future practitioners.

G. Y. Craig, D. B. McIntyre, and C. D. Waterston, *James Hutton's Theory of the Earth: The Lost Drawings* (Edinburgh: Scottish Academic Press Ltd., 1978). These drawings are simply superb. They must be seen to be appreciated.

Chapter 3

Daniel Defoe, *The Life and Strange Adventures of Robinson Crusoe* in *The Works of Daniel Defoe* (New York: Thomas Y. Crowell and Company, 1903), pp. 172, 178.

Thomas S. Kuhn, *The Structure of Scientific Revolutions,* 2d ed. (Chicago: University of Chicago Press, 1970). Kuhn's ideas have been the center of discussion for over a decade because scientific revolutions are not all exactly alike.

Louis Agassiz, *Studies on Glaciers,* preceded by the *Discourse of Neuchatel,* translated and edited by Albert V. Carozzi (New York: Hafner Publishing Co., 1967). All of the quotations on glacial features, including those about Jean de Charpentier, were taken from this book: pp. xv, xvi, lxvii, and his knowledge in context.

Henry Wadsworth Longfellow, *The Poetical Works* (Boston: Houghton Mifflin and Co., 1886), vol. 3, p. 245. The Agassiz poem contains one other beautiful couplet that should appeal to backpackers the world over:

> And the rush of mountain streams
> From glaciers clear and cold

John W. Harrington, *Discovering Science* (Boston: Houghton Mifflin, 1981), pp. 45–59. Material on both Charles Darwin and Alfred Wegener has been rewritten from this earlier book of mine.

Sir Gavin de Beer, *Charles Darwin* (Garden City, New York: Doubleday, 1965), pp. 30, 34, 82, and 132.

Sir Charles Lyell, *Principles of Geology*, vol. 1 (London: John Murray, 1830). Lyell's awareness of changes in animals and plants through geologic time may surprise some readers who may not have known of the widespread interest in organic evolution that predated Darwin's work.

Charles Darwin, *On the Origin of Species* (London: John Murray, 1859).

Charles Darwin, *Journal of Researches of the Voyage of the Beagle* (New York: P. F. Collier, 1902), p. 429. This is a reprint of the famous 1845 edition.

Frank H. T. Rhodes, and Richard O. Stone, eds. *Language of the Earth* (New York: Pergamon Press, 1981), p. 44.

Alfred L. Wegener, *The Origin of Continents and Oceans,* 3d ed., trans. J.F.A. Skerl (London: Methuen, 1924).

Arthur Holmes, *Principles of Physical Geology,* 2d ed. (New York: The Ronald Press Co., 1965). Figures 539 and 540 from p. 736 were used as the source for our figure 3-8.

Percy Bysshe Shelley in Curtis Hidden Page, ed., *British Poets of the Nineteenth Century,* 4th ed. (New York: B. H. Sanborn & Co., 1910), pp 288–91. The entire poem is well worth reading for its technical insights alone.

Chapter 4

Jan Myrdal, *Report from a Chinese Village,* trans. (from Swedish) Maurice Michael (New York: Pantheon Books, 1975), p. xvii. Mau Ke-yeh's remark was taken from the special introduction to this edition. It's amazing how things fit together; random reading in one field introduces so many ideas and viewpoints that can be transposed and used in another.

Part II

David Macauly. "Great Moments in Architecture," *Atlantic Monthly* 240 (December 1977):71. Used by permission of the publisher. Other cartoons in this article are equally appealing.

Chapter 5

George L. Kittredge in Houston Peterson, ed., *Great Teachers Portrayed by Those Who Studied Under Them* (New Brunswick, N. J., Rutgers University Press, 1966), p. 225.

Claude C. Albritton, Jr., *The Abyss of Time* (San Francisco: Freeman Cooper and Co., 1980), pp. 20–41. Chapters 2 and 3 are devoted to the work of Nicolaus Steno. We have drawn heavily from the ideas in this text.

John Garrett Winter, *Contributions to the History of Science, Part II: The Prodromus of Nicolaus Steno's Dissertation Concerning a Solid Body Enclosed by the Process of Nature Within a Solid* (Ann Arbor, Michigan: University of Michigan Press, 1930). Think of Steno as a great teacher as well as a great scientist.

Stephen Jay Gould, "The Titular Bishop of Titiopolis," *Natural History* 90 (May 1981): 20–24. This is an unusual analysis of Steno's style as a thinker. The title refers to his duties as a priest after he gave up geology. A titular bishop was responsible for presiding over pagan lands that were not available for actual residence.

Norma Lorre Goodrich, *The Medieval Myths: Beowulf and the Fiend Grendel* (New York: The New American Library, 1961), pp. 17–19.

Percy Bysshe Shelley, in Curtis Hidden Page, ed., *British Poets of the Nineteenth Century*, 4th ed. (New York: B. H. Sanborn & Co., 1910).

Robert Frost, *The Poetry of Robert Frost*, ed. Edward Connery Latham (New York: Holt, Rinehart and Winston, 1969). Copyright 1967 by Lesley Frost Ballentine. Used by permission of Holt, Rinehart and Winston, Inc. and the estate of Robert Frost, p. 250.

Chapter 6

Rudyard Kipling, "The Lost Legion," *Ballads and Barrack Room Ballads*, new ed. (New York: Macmillan Co., 1893), pp. 74–76. Anyone who is devoted to the swaggering romantics of the nineteenth century will thrill at this wonderful poem.

Joseph Needham, with the research assistance of Wang Ling, *Mathematics and the Sciences of Heaven and Earth* (Cambridge: at the University Press, 1970), (vol. 3 of *Science and Civilization in China*). The quotation from Shen Kua was taken from p. 614, and that from Chu Hsi from p. 598. Used by permission of the Cambridge University Press.

Henry Kiepert, *Atlas Antiquitus: Twelve Maps of the Ancient World for Schools and Colleges*, 11th ed. (Boston: Leach, Shewell and Sanborn, 1902). This was the source of figure 6-1.

Amadeus W. Grabau, *Stratigraphy of China, Part I, Paleozoic and Older* (Peking: Ministry Agriculture and Commerce, Geological Survey, 1923–24). Chu Hsi's comments were taken from the introductory material.

George Rawlinson, trans., and Manuel Komeroff, ed., *The History of Herodotus* (New York: Tudor Publishing Co., 1928). Used by permission of Tudor Publishing Co. This is the book that contains the great line, "Thus fought the Greeks at Thermopylae."

H. E. Wulff, "Qanats of Iran," *Scientific American* 218 (April 1968):94–105.

John W. Harrington, *Discovering Science* (Boston: Houghton Mifflin Co., 1981). My earlier book, the source of figure 6-7, contains a more detailed discussion of the life and work of Eratosthenes. Purists interested in a scholarly study that develops a different concept of the measurements made by Eratosthenes and his colleagues should consult Dennis Rawlings, "Eratosthenes' Geodesy Unraveled: Was There A High-Accuracy Hellenistic Astronomy?" *Isis* 73 (June 1982): 259–65.

J. Oliver Thompson, *History of Ancient Geography* (New York: Biblo and Tannen, 1965). This book is an important compilation of information on the geographers of the ancient world. Many of them have made contributions to our present world view and yet, like Pytheas, have been almost forgotten or, like Eratosthenes, slighted. Scholars will find pp. 143–51 particularly rewarding reading.

Matthew Fontaine Maury, *The Physical Geography of the Sea and Its Meteorology*, ed. John Leighly (Cambridge, Mass.: The Belknap Press of Harvard University Press, 1963). I grew up in Richmond, Virginia. There is a monument to Maury on Monument Avenue sculpted by the father of my boy scout camp counselor. Heroes are important to young dreamers. This monument was and still is an inspiration to me. Maury will always be there, in my mind, contemplating the whole of the earth as he holds it in his hands.

Chapter 7

Colin Fletcher, *The Man Who Walked Through Time* (New York: Vintage Books, 1967), p. 6.

Grove Karl Gilbert, *Geology of the Henry Mountains* (Washington, D. C.: Government Printing Office, 1880). Geological reports no longer have the joys of research and field adventures that Grove Karl Gilbert put into his humanistic accounts. He thought it was important to define the route to the Henry Mountains that would show off the Indian pictographs to their best advantage; besides, there was water in this canyon. No desert rat takes sweet water for granted.

Murray Macgregor, *Excursion Guide to the Geology of Arran* (Glasgow: The Geological Society of Glasgow and the University of Glasgow, 1965), p. 125. The quotation from James Hutton was assembled by Murray Macgregor from two earlier sources.

John Playfair, "The Life of Dr. Hutton," *Transactions of the Royal Society of Edinburgh* 5, part 3(1805):39–99. The most readily available source of this description of the day at Siccar Point is in a reprint edited by George W. White, *Contributions to the History of Geology*, vol. 5 (Darien, Conn.: Hafner Publishing Company, 1970), pp. 175–77.

Chapter 8

J. Lawrence Kulp, "The Geological Time Scale," *Science* 133 (1961):1105–14.

John G. C. M. Fuller, "The Industrial Basis of Stratigraphy: John Strachey, 1671–1743, and William Smith, 1769–1839." *American Association of Petroleum Geologists Bulletin* 53 (1968). Fuller has written one of those rare articles in the scientific literature that establishes the character of the people as well as their work.

Thomas Mann, *The Magic Mountain* (New York: Alfred A. Knopf, 1938), p. 287. Mann's point was that we place artificial boundaries on the flow of time.

Sir William Thompson, Lord Kelvin, *The Age of the Earth as an Abode Fitted for Life,* Annual Report of the Board of Regents, Smithsonian Institution (Washington, D. C.: Government Printing Office, July 1897, 1898), pp. 340–41. This is Kelvin's annual address of the Victoria Institute with later additions as reprinted by the Smithsonian Institution. There are many other interesting quotations in this report, pp. 337–57.

Charles Darwin, *On the Origin of Species* (Cambridge, Mass.: Harvard University Press, 1964), pp. 282–87 (a facsimile of the first edition, 1859, with an introduction by Ernst Mayr). These quotations were included because Lord Kelvin thought they were important enough to quote in his article.

W. B. Harland, A. G. Smith, and B. Wilcock, "The Phanerozoic Time Scale" (Symposium dedicated to Professor Arthur Holmes) *Quarterly Journal of the Geological Society of London* 120s (1964). Case no. 1, the Murray Shale, was taken from catalog item 183, p. 375; case no. 2, the Palisade sill, is catalog item 9, p. 277; case no. 3, the mica-bearing volcanic ash, is catalog item 27, p. 291.

Chapter 9

Fred C. Kelley, *Miracle at Kitty Hawk* (New York: Farrar, Straus & Giroux, 1951). The quotation was taken from a letter dated June 7, 1903, six months and ten days before their first powered flight. It is found on p. 91.

Gudmundur Palmason and Kristjan Saemundsson, "Iceland in Relation to the Mid-Atlantic Ridge," *Annual Review of Earth and Planetary Science* 2 (1974):25–50

Frederick J. Sawkins, Clement G. Chase, David G. Darby, and George Rapp, Jr., *The Evolving Earth* (New York: Macmillan Publishing Co., and London: Collier Macmillan Publishers, 1978), p. 163.

M. Barazangi and J. Dorman, "World Seismicity Maps Compiled from ESSA, Coast and Geodetic Survey Epicenter Data, 1961–1967," *Seismological Society of America Bulletin* 59 (1969):385–400.

O. F. Morshead, ed., *Everybody's Pepys, the Diary of Samuel Pepys, 1660–1669* (New York: Harcourt Brace Jovanovich, 1926), pp. 93–94. Pepys was not a scientist himself, but he must have had a level of curiosity that gained the respect of the scientific community of the time. The flyleaf of Isaac Newton's revolutionary book, *Principia Mathematica,* published in 1687, contained these words: "I Newton, He wrote it. S. Pepys, He published it."

Sir Jonas Moore, *A True and Natural Description of the Great Level of the Fenns* (no standard bibliographic data, British Museum Library, London, 1685). Retrieving this book from the depths of the British Museum Library was an interesting experience. I was amazed to find it was a little thing, bound in green cloth, with the poem at the end.

Walter Sullivan, *Continents in Motion* (New York: McGraw-Hill Book Co., 1974), p. 57, contains the story of Bruce Heezen's talk at Princeton.

Harry H. Hess, "History of Ocean Basins," in *Petrologic Studies.* A volume in Honor of A. F. Buddington, A. E. J. Engel, Harold L. James, and B. F. Leonard, eds., Special Volume of the Geological Society of America, 1962.

Robert S. Dietz, "Continents and Ocean Basin Evolution by Spreading of the Sea Floor," *Nature* 190 (1961):854–57.

Mason L. Hill and T. W. Dibblee, Jr., "San Andreas, Garlock and Big Pine Faults, California: A Study of the Character, History and Tectonic Significance of their Displacements," *Geological Society of America Bulletin* 64 (1961):443–58.

Vincent Matthews III, "Correlation of Pinnacles and Neenach Volcanic Formations and Their Bearing on San Andreas Fault Problem," *American Association of Petroleum Geologists Bulletin* 60 (1976):2128–41.

Hugo Benioff, "Orogenesis and Deep Crustal Structure—Additional Evidence from Seismology," *Geological Society of America Bulletin* 65(1954):385–400. This was a key paper that convinced many geologists that oceanic crust was plunging beneath the edge of island arcs.

Kirtley F. Mather and Shirley L. Mason, *A Source Book in Geology* (New York: McGraw-Hill Book Co., 1939): pp. 406–15.

David Johnston's remarks and those of his father were taken from "Mount Saint Helens Diary, A Sunday Holocaust," Vancouver, Washington, The Columbian, Inc., 28 May, 1980.

Index

Page numbers in *italics* refer to figures in the text.

The author would like to thank the following for their permission to reprint:

James Hutton's Theory of The Earth: The Lost Drawings. Scottish Academic Press, Edinburgh, Scotland, 1978.

Studies On Glaciers by Louis Agassiz, translated and edited by Albert V. Carozzi (Copyright © 1967 by Hafner Publishing Company).

"Early Work on the Grand Canyon," from *Great Moments in Architecture* by David Macaulay. Copyright © 1978 by David Macaulay. Reprinted by permission of Houghton Mifflin Company.

"Once by the Pacific," from *The Poetry of Robert Frost* edited by Edward Connery Lathem. Copyright 1928, © 1969 by Holt, Rinehart and Winston. Copyright © 1956 by Robert Frost. Reprinted by permission of Holt, Rinehart and Winston, Publishers.

From *Markings* by Dag Hammarskjold, translated by Leif Sjoberg and W. H. Auden. Copyright © 1966 by Dag Hammarskjold. Reprinted by permission of Alfred A. Knopf, Inc.

Manuel Komroff, *The History of Herodotus* translated by George Rawlinson, New York: Tudor Publishing Co., 1928.

Science and Civilization in China Vol. III by Joseph Needham, Cambridge University Press, 1979.

"Concerning the System of the Earth" by James Hutton (Vol. 5, *Contributions to the History of Geology* edited by George W. White) Intro. by Victor A. Eyles, © Macmillan

2834